CELL BIOLOGY RESEARCH PROGRESS

CADHERINS

TYPES, STRUCTURE AND FUNCTIONS

CELL BIOLOGY RESEARCH PROGRESS

Additional books and e-books in this series can be found on Nova's website under the Series tab.

Cell Biology Research Progress

CADHERINS

TYPES, STRUCTURE AND FUNCTIONS

JONATHAN MCWILLIAM
EDITOR

Copyright © 2020 by Nova Science Publishers, Inc.

All rights reserved. No part of this book may be reproduced, stored in a retrieval system or transmitted in any form or by any means: electronic, electrostatic, magnetic, tape, mechanical photocopying, recording or otherwise without the written permission of the Publisher.

We have partnered with Copyright Clearance Center to make it easy for you to obtain permissions to reuse content from this publication. Simply navigate to this publication's page on Nova's website and locate the "Get Permission" button below the title description. This button is linked directly to the title's permission page on copyright.com. Alternatively, you can visit copyright.com and search by title, ISBN, or ISSN.

For further questions about using the service on copyright.com, please contact:
Copyright Clearance Center
Phone: +1-(978) 750-8400   Fax: +1-(978) 750-4470   E-mail: info@copyright.com.

NOTICE TO THE READER

The Publisher has taken reasonable care in the preparation of this book, but makes no expressed or implied warranty of any kind and assumes no responsibility for any errors or omissions. No liability is assumed for incidental or consequential damages in connection with or arising out of information contained in this book. The Publisher shall not be liable for any special, consequential, or exemplary damages resulting, in whole or in part, from the readers' use of, or reliance upon, this material. Any parts of this book based on government reports are so indicated and copyright is claimed for those parts to the extent applicable to compilations of such works.

Independent verification should be sought for any data, advice or recommendations contained in this book. In addition, no responsibility is assumed by the Publisher for any injury and/or damage to persons or property arising from any methods, products, instructions, ideas or otherwise contained in this publication.

This publication is designed to provide accurate and authoritative information with regard to the subject matter covered herein. It is sold with the clear understanding that the Publisher is not engaged in rendering legal or any other professional services. If legal or any other expert assistance is required, the services of a competent person should be sought. FROM A DECLARATION OF PARTICIPANTS JOINTLY ADOPTED BY A COMMITTEE OF THE AMERICAN BAR ASSOCIATION AND A COMMITTEE OF PUBLISHERS.

Additional color graphics may be available in the e-book version of this book.

Library of Congress Cataloging-in-Publication Data

Names: McWilliam, Jonathan, editor. Title: Cadherins : types, structure and functions / [edited by] Jonathan McWilliam.
Description: Hauppauge : Nova Science Publishers, [2020] | Series: Cell biology research progress | Includes bibliographical references and index. | Summary: "Cadherins are adhesion molecules which mediate homophilic Ca2+-dependent intercellular adhesion. They play a crucial role in tissue morphogenesis during embryonic development and in the maintenance of tissue architecture in adults. In this compilation, the authors review and discuss the structure of the cadherin-mediated junctions in seminiferous tubules and their functions in male fertility. Additionally, role of E-cadh in uterine and mammary gland homeostasis is reviewed, and the disruptive patterns of E-cadh expression in neoplastic conditions of the uterus and mammary glands in humans and domestic dogs and cats are described"-- Provided by publisher.
Identifiers: LCCN 2020024377 (print) | LCCN 2020024378 (ebook) | ISBN 9781536180770 (paperback) | ISBN 9781536180961 (adobe pdf)
Subjects: LCSH: Cadherins. | Cell adhesion molecules.
Classification: LCC QP552.C42 C34 2020 (print) | LCC QP552.C42 (ebook) | DDC 572/.4--dc23
LC record available at https://lccn.loc.gov/2020024377
LC ebook record available at https://lccn.loc.gov/2020024378

Published by Nova Science Publishers, Inc. † New York

CONTENTS

Preface		**vii**
Chapter 1	Standing Out from the Crowd: Functions of T-Cadherin in Health and Disease *Maria Philippova and Therese J. Resink*	**1**
Chapter 2	Cadherin-Mediated Cell Adhesion within the Seminiferous Tubules *Rita Payan-Carreira and Dario Santos*	**95**
Chapter 3	Disruption of E-Cadherin Pattern in Uterine and Mammary Tumours *Adelina Gama, Fernanda Seixas, Maria dos Anjos Pires, Fernando Schmitt and Rita Payan-Carreira*	**129**
Index		**163**

PREFACE

Cadherins are adhesion molecules which mediate homophilic Ca2+-dependent intercellular adhesion. They play a crucial role in tissue morphogenesis during embryonic development and in the maintenance of tissue architecture in adults.

In this compilation, the authors review and discuss the structure of the cadherin-mediated junctions in seminiferous tubules and their functions in male fertility.

Additionally, role of E-cadh in uterine and mammary gland homeostasis is reviewed, and the disruptive patterns of E-cadh expression in neoplastic conditions of the uterus and mammary glands in humans and domestic dogs and cats are described. (Imprint: Nova)

Chapter 1 - Cadherins are adhesion molecules which mediate homophilic Ca^{2+}-dependent intercellular adhesion. They play a crucial role in tissue morphogenesis during embryonic development and in the maintenance of tissue architecture in the adult. Cadherin-based adherens junctions and downstream signaling pathways are important for regulation of many processes during tissue remodeling such as cell sorting, polarity, migration, differentiation and survival. The cadherin superfamily is heterogeneous. Apart from the main large sub-families of structurally related members, it includes several atypical cadherins and cadherin-related proteins with unique molecular structures. Among these is T-cadherin (cadherin-13) which lacks transmembrane and cytosolic domains and is anchored to the plasma membrane *via* a glycosylphosphatidylinositol anchor. Due to the absence of an intracellular molecular moiety, T-cadherin-based intercellular contacts are weak and T-cadherin-dependent cellular processes are mediated by signaling mechanisms that are entirely different from those utilized by classical transmembrane cadherins. T-cadherin was originally described in the embryonic nervous system where it functions as a guidance molecule navigating projecting motor axons. Subsequently T-cadherin was demonstrated to regulate cell motility, proliferation, survival and phenotype of vascular endothelial and smooth muscle cells, cardiomyocytes, keratinocytes and several cancer cell types. Of particular interest is its function in angiogenesis and also its involvement in cardioprotective effects of adiponectin, an adipose tissue-derived hormone which regulates glucose and fatty acid metabolism and supresses progression of atherosclerosis. In the current chapter the authors review the existing knowledge and recent studies on T-cadherin structure, signaling, functions in different tissues, and relevance to pathogenesis of neurological disorders, cardiovascular disease and cancer.

Chapter 2 - Cadherins (Cadh) are key-molecules in Adherens junctions (AJs). They are multiprotein complexes mediating cell-cell adhesion, and particularly important to shape cell polarity, provide plasticity and maintain architectural integrity. Cadh, a large superfamily of cell surface glycoproteins, present a unique extracellular region domain folding like the immunoglobulin domains. They are found in a wide array of species and a multitude of tissues, including the testis. In the mammalian testis, the seminiferous tubules represent a unique type of epithelium-like tissue, composed of two different cellular populations: the Sertoli somatic cells and the spermatogenic cells. Different sorts of cell-to-cell attachments connect adjacent Sertoli cells and Sertoli to germ cells. The overall arrangement of junctions forms the blood-testis barrier. These connections offer an immune-privileged environment to the developing germ cells, and the nutritional and metabolic support to germ cells while offering particular plasticity to the tubular structure. They allow the migration of differentiated germ cells from the basal towards the adluminal compartment while providing a tight-fitting barrier for paracellular translocation of molecules and particles. Between adjacent Sertoli cells, various types of homotypic adherens junctions exists, while heterotypic junctions are present between Sertoli and spermatogonia (basolateral junctions) or spermatid heads (apical junctions). Intercellular N-cadherin connections, anchored in cytoplasmic plaques involving (but not limited to) actin filaments, form different morphological types of AJs. All sorts of AJs work together with tight and gap junctions to form the blood-testis barrier. The integrity of the different adherens junctions are critical for the spermatogenic process and the production of viable spermatozoa. In this chapter, the authors propose to review and discuss the structure of the cadherin-mediated junctions in the seminiferous tubules and their function in male fertility.

Chapter 3 - E-cadherin (E-cadh), a member of the classic cadherins superfamily, plays an important role in epithelial cell-to-cell adhesion, encompassing the dynamic interactions between adjacent cells including the control of morphogenesis, maintenance of cell polarity and tissue architecture. Cadherins comprise a large family of cell surface glycoproteins, presenting unique extracellular regions domains known as the cadherin motifs or domains, which fold like immunoglobulin domains. They mediate strong Ca^{2+}-dependent homophilic interactions between neighbouring epithelial cells, resulting in the formation of cell adhesion "zippers." E-cadh cytoplasmic tail links to catenins and, thereby to the actin cytoskeleton and signalling proteins to form a cell-cell signalling centre: it regulates several intracellular signal transduction pathways, including Wnt/β-catenin, PI3K/Akt, Rho GTPase, and NF-KB signalling. E-cadh plays a crucial role in the barrier formation of polarized epithelial cell layers at the interfaces contacting with the external environment, namely the uterus and the mammary gland. The maintenance of these barriers could be considered as a prime immunologic function of E-cadh, compartmentalizing potentially harmful agents away from the underlying tissue. Disruption of classical cadherin expression has been related to the occurrence of diseases driving disturbances in tissue architecture, such as inflammation and cancer. In cancer, loss of E-cadh expression/function increases cell proliferation, cell migration, and disruption of epithelial cell homeostasis, driving cell dissociation and scattering. Alterations of E-cadh expression have also been reported during particular moments of the female reproductive physiology, namely throughout the oestrous cycle or during the embryo-maternal interaction at embryo implantation (early pregnancy). Several studies have shown the downregulation of E-cadh in malignant epithelial tumours, which has been associated with loss of cell differentiation, epithelial to mesenchymal transition (EMT) and invasion. Data also suggest that loss of E-cadh may be associated with malignant progression, metastasis, and reduced survival in multiple cancer patients. In this chapter, the authors review and discuss the role of E-cadh in the uterine and mammary gland homeostasis and describe disruptive patterns of E-cadh expression in neoplastic conditions of the uterus and mammary glands in human and domestic dogs and cats.

In: Cadherins
Editor: Jonathan McWilliam
ISBN: 978-1-53618-077-0
© 2020 Nova Science Publishers, Inc.

Chapter 1

STANDING OUT FROM THE CROWD: FUNCTIONS OF T-CADHERIN IN HEALTH AND DISEASE

Maria Philippova[1],, PhD and Therese J. Resink[2], PhD*

[1]Tissue Engineering Laboratory, Departments of Biomedicine and Surgery, Basel University Hospital, Switzerland
[2]Signal Transduction Laboratory, Department of Biomedicine, Basel University Hospital, Switzerland

ABSTRACT

Cadherins are adhesion molecules which mediate homophilic Ca^{2+}-dependent intercellular adhesion. They play a crucial role in tissue morphogenesis during embryonic development and in the maintenance of tissue architecture in the adult. Cadherin-based adherens junctions and downstream signaling pathways are important for regulation of many processes during tissue remodeling such as cell sorting, polarity, migration, differentiation and survival. The cadherin superfamily is heterogeneous. Apart from the main large sub-families of structurally

---

* Corresponding Author's E-mail: maria.filippova@unibas.ch.

related members, it includes several atypical cadherins and cadherin-related proteins with unique molecular structures. Among these is T-cadherin (cadherin-13) which lacks transmembrane and cytosolic domains and is anchored to the plasma membrane *via* a glycosylphosphatidylinositol anchor. Due to the absence of an intracellular molecular moiety, T-cadherin-based intercellular contacts are weak and T-cadherin-dependent cellular processes are mediated by signaling mechanisms that are entirely different from those utilized by classical transmembrane cadherins. T-cadherin was originally described in the embryonic nervous system where it functions as a guidance molecule navigating projecting motor axons. Subsequently T-cadherin was demonstrated to regulate cell motility, proliferation, survival and phenotype of vascular endothelial and smooth muscle cells, cardiomyocytes, keratinocytes and several cancer cell types. Of particular interest is its function in angiogenesis and also its involvement in cardioprotective effects of adiponectin, an adipose tissue-derived hormone which regulates glucose and fatty acid metabolism and supresses progression of atherosclerosis. In the current chapter we review the existing knowledge and recent studies on T-cadherin structure, signaling, functions in different tissues, and relevance to pathogenesis of neurological disorders, cardiovascular disease and cancer.

ABBREVIATIONS

ADHD	attention deficit hyperactivity disorder
ADIPOQ gene	adiponectin gene
AdipoQ$^{-/-}$ mice	adiponectin knockout mice
Akt	AKT8 virus oncogene cellular homolog/protein kinase B
5-ALA	5-aminolevulinic acid
AMPK	AMP-activated protein kinase
APN	adiponectin
ApoE	apolipoprotein E
ASD	autism spectrum disorder
α-SMA	alpha-smooth muscle actin
BAL	bronchoalveolar lavage
BCC	basal cell carcinoma
CHD13 gene	cadherin-13/T-cadherin gene

Cdh13⁻/⁻ mice	T-cadherin–knockout mice
CML	chronic myeloid leukaemia
COPD	chronic obstructive pulmonary disease
CRC	colorectal cancer
CVD	cardiovascular diseases
EC	endothelial cell
EC repeats	extracellular cadherin (EC) repeats
EGF	epidermal growth factor
EGFR	epidermal growth factor receptor
EMT	epithelial-to-mesenchymal transition
eNOS	nitric oxide synthase
ER	estrogen receptor
ERK1/2	extracellular signal-regulated kinase 1/2
GPI	glycosylphosphatidylinositol
GSK3β	glycogen synthase kinase 3β
GWAS	genome-wide association studies
HB-EGF	heparin-binding epidermal growth factor-like growth factor
HCC	hepatocellular carcinoma
HEK293	human embryonic kidney
IGF-1R	insulin growth factor-1 receptor
IGF	insulin like growth factor
IRS-1	insulin receptor substrate 1
LDL	low density lipoprotein
LOH	loss of heterozygosity
MAPK	mitogen-activated protein kinase
MEK	MAPK/Erk kinase
MMTV-PyV-mT model	mouse mammary tumor virus-polyoma virus middle transgenic model
MPs	microparticles
mTOR	mammalian target of rapamycin
NSCLC	non-small cell lung cancer
OSS	oral squamous carcinoma

p38 MAPK	p38 mitogen activated protein kinase
PCa	prostate cancer
PDGF	platelet-derived growth factor
PERK	protein kinase RNA-like endoplasmic reticulum kinase
PI3K	phosphoinositide 3-kinase
PpIX	protoporphyrin IX
Rac	Ras-related C3 botulinum toxin substrate
Rac1	Ras-related C3 botulinum toxin substrate 1
RhoA	Ras homolog family member A
S6K1	(p70) S6 kinase 1
SCC	squamous cell carcinoma
SMC	smooth muscle cell
TAC	transverse aortic constriction
TNFα	tumour necrosis factor alpha
VEGF	vascular endothelial growth factor
WT	wild type

1. INTRODUCTION

1.1. Cadherin Superfamily of Intercellular Adhesion Molecules

Cadherins play essential roles in important morphogenetic and differentiation processes during development, and in maintaining tissue integrity and homeostasis in adults. Cadherins mediate Ca^{2+}-dependent trans-junctional homophilic interactions at cell-cell interfaces that function to establish strong cell-cell adhesion and to define adhesive specificities of cells. The functions of cadherins extend beyond mere establishment of intercellular adhesion to multiple aspects of tissue organization and morphogenesis, including cell recognition and sorting, boundary formation in tissues, induction and maintenance of structural and functional cell and tissue polarity, cytoskeletal organization, cellular phenotype modulation, cell migration, cell proliferation and cell survival.

The cadherin superfamily is large and heterogeneous. It comprises more than 350 members found in various species; more than 100 members have been described in vertebrates [1]. According to the classification of Hulpiau and van Roy [2], which is based on an in-depth review and analysis of amassed data on molecular evolution and genomic sequencing, the superfamily is divided into two main branches, namely a cadherin major branch and a cadherin-related major branch. The cadherin major branch is comprised of two families C-1 (with type-I, type-II, desmocollins, desmogleins, 7D and solitary cadherin subfamilies) and C-2 (with Flamingo, type III and type IV cadherin subfamilies). The cadherin-related branch is subdivided into in four families: Cr-1a (protocadherins), Cr-1b (RET, FAT, Dachsous), Cr-2 (CDHR and μ-CDHR) and Cr-3 (FAT-like, CDHR28, CDHR15 and calsyntenins). The foremost common structural feature of cadherins is the presence of Ca^{2+}-binding extracellular cadherin (EC) repeats in their ectodomain, the number of these repeats varying from two in calsyntenins to 34 in FAT-like cadherins. While there are several conserved motifs in subsets of cytoplasmic domains, these domains are even more diverse than ectodomains [2]. Readers are referred to a range of excellent recent articles that variously review the evolution of cadherin family members, their diverse functions in tissue development and maintaining tissue integrity, and their many molecular mechanisms of action [3-28].

1.2. Distinguishing T-Cadherin from Other Cadherin Family Members: Structural Features

T-cadherin (cadherin-13, H-cadherin), which is encoded by the *CDH13* gene that is located on the chromosome 16q24 region [29], is a solitary subfamily member of the C-1 cadherin family [2]. T-cadherin is an atypical cadherin with several unique structural and functional characteristics. Its foremost distinctive feature is that it lacks transmembrane and cytoplasmic domains and is instead tethered to the plasma membrane by a glycosylphosphatidylinositol (GPI) anchor [30].

Another unique feature of T-cadherin is the structure of its relatively small adhesive pocket on EC1 domain, which has an isoleucine residue at position 2 instead of the conserved tryptophan residue in type I and type II cadherins [31]. Due to this amino acid substitution T-cadherin cannot mediate intercellular adhesion by formation of dimers through N-terminal EC1 domain interactions and "strand swapping" between partner molecules; instead T-cadherin forms X-dimer binding intermediates through an alternative non-swapped interface near EC1-EC2 Ca^{2+}-binding sites [15, 32]. T-cadherin has only a low Ca^{2+}-dependent homophilic-based adhesive capacity [31-33], and formation of mechanically stable cell-cell contacts is limited by the absence of cytoplasmic domain anchorage *via* armadillo catenins to the cytoskeleton. Moreover, T-cadherin is distributed diffusely over the cell surface with only mild enrichment in intercellular adherens junctions of quiescent cell monolayers, is primarily located on the apical surface of polarized cells, and undergoes redistribution to the leading edge of migrating cells [33-35]. Therefore, T-cadherin homophilic interactions are functionally more appropriate to dynamic adhesion-deadhesion processes than to strong intercellular adhesion.

1.3. Distinguishing T-Cadherin from Other Cadherin Family Members: Signaling Partners and Heterophilic Ligands

Another distinctive feature of T-cadherin concerns the mechanisms whereby it may trigger signal transduction. Other C-1 cadherin family members initiate signaling in the cytoplasm and nucleus through engagement of the intracellular domain with an assortment of intracellular binding partners (e.g., cytoskeletal regulators, protein kinases/ phosphatases, transcriptional cofactors) and lateral interactions of the transmembrane domain with growth factor receptors and other plasma membrane-located signaling molecules. However, due to its lack of transmembrane and cytoplasmic domains T-cadherin cannot trigger intracellular signaling *via* such direct engagement mechanisms. As demonstrated for other GPI-anchored proteins [36], T-cadherin-dependent signal transduction necessarily requires some lateral interaction with accessory membrane molecular adaptors together with spatio-temporal reorganization of lipid rafts (where T-cadherin locates [37, 38]) and/or changes in T-cadherin protein clustering at the cell surface [39]. Investigations on molecular adaptors for T-cadherin are rare and restricted to endothelial cells (ECs): adaptors identified include insulin receptor, stress chaperone GRP78/BiP, integrin-linked kinase, GABA-A receptor α1 subunit, integrin β3, collagen α2 (I) chain and two hypothetical proteins, LOC124245 and FLJ32070 [40-42]. A fourth important distinctive property of T-cadherin relates to heterophilic interactions: it is the only member of the cadherin family for which a capacity for heterologous binding with any circulating or secreted ligand has been reported. Ligands identified to date include native low density lipoprotein (LDL) [43-46] and adiponectin (APN/ACRP30), an adipocyte-derived adipocytokine which is encoded by the *ADIPOQ* gene [47, 48]. Confocal microscopy, FRAP and FRET imaging *in vitro* using HEK293 cell line demonstrated that their respective ligation to T-cadherin occurs *via* distinct mechanisms. Native LDL binds to both 130 kDa prepro and 105 kDa mature T-cadherin proteins [45] and the GPI moiety of T-cadherin is necessary and sufficient for this interaction [49]. APN binds preferentially to 130 kDa prepro-T-cadherin protein and requires the region encompassing EC1-EC2 domains of T-cadherin [48], overlapping the region reported for homophilic *trans* interaction of T-cadherin [32]. Both LDL and APN binding to T-cadherin cause cell surface spatiotemporal organization of T-cadherin but in different modalities [39]. LDL induces rapid formation of short-lived T-cadherin clusters for which the presence of membrane cholesterol as well as an intact actin cytoskeleton are obligate, whereas T-cadherin cluster formation induced by APN is stable, independent of cholesterol, sensitive to actin perturbation and accompanied by internalization of T-cadherin [39].

Biochemical ligand-blotting techniques have identified LDL-T-cadherin binding interactions [44], but specific colocalization of LDL to T-cadherin in tissues/cell types *in vivo* has yet to be demonstrated. Direct physical association between T-cadherin and high molecular weight APN has been demonstrated by co-immunoprecipitation experimentation using ECs, C2C12 murine myotubes and human embryonic kidney (HEK) cells [47, 50, 51]. APN colocalizes with cell surface expressed T-cadherin *in vivo* in a variety of tissues including the vasculature (on ECs and smooth muscle cells (SMCs) [35, 50, 52, 53], heart (on cardiomyocytes) [51, 52] and skeletal muscle (on myocytes, capillaries and larger blood vessels) [50, 52, 54]. Investigations in T-cadherin–knockout ($Cdh13^{-/-}$) mice have established that T-cadherin plays a crucial role in tissue accumulation of APN [35, 50-52, 54]. The pathophysiological relevance of interactions between T-cadherin and its heterophilic ligands in various tissues is discussed below in respective sections.

$Cdh13^{-/-}$ mice live well into adulthood without overt pathological phenotypes [35, 51], and identification of functions for T-cadherin *in vivo* have required challenging the mice in disease models. In the following we appraise current *in vitro* and *in vivo* knowledge regarding on T-cadherin expression and functions in different tissues and cell types and how it may participate in progression of many pathological conditions including cardiovascular diseases (CVD), neurological disorders and cancers.

2. THE NERVOUS SYSTEM

2.1. Axon Guidance

The development of functional neural circuits in the nervous system during embryogenesis and early postnatal period depends on a complex series of events involving axon guidance, axon and dendrite arborization, formation and maturation of synaptic structures, and elimination of unnecessary neural connections. Aberrant spatial and temporal orchestration of axon guidance processes have been linked to many neurological disorders [55]. Several types of structurally diverse molecules have been attributed a role in axon navigation and target recognition, among them semaphorins, netrins, slits, ephrins, cadherins, *inter alia*. The pioneering works of the group of Barbara Ranscht have not only identified

T-cadherin as a novel guidance molecule regulating pathfinding of projecting axons but also laid the foundation for the understanding of the principles underlying T-cadherin effects in other tissues.

In the chick embryo T-cadherin was suggested to be among the molecular cues which define migration routes of neural crest cells through somite-derived sclerotomes and maintain somite polarity. The assumption was based on the observation for differential expression of T-cadherin along the rostrocaudal axis of each sclerotome that coincided in time with the invasion of the first neural crest cells into the rostral parts of the somite and was increased towards the caudal parts of somites avoided by neural crest cells and extending axons [56]. Similarly, in the developing chick hindlimb muscle projecting motor axons avoided regions positive for T-cadherin expression which dynamically changed during initial extension of axons from the spinal cord, sorting into different nerve trunks and formation of terminal synapses in the muscle [57]. These data suggested that T-cadherin acts as a negative guidance cue to define trajectories of migrating neural crest cells and projecting axons. Direct evidence for this hypothesis came from *in vitro* studies which demonstrated that both T-cadherin substrata and soluble recombinant T-cadherin protein inhibited neurite growth through homophilic adhesion-dependent mechanism [58]. Immunohistochemical and knockout studies confirmed that T-cadherin contributes to axonal pathfinding by cortical projection neurons [59].

T-cadherin is likely to act in conjunction with other adhesion molecules to orchestrate formation of neural circuits. In the chick embryo T-cadherin and classical type I N-cadherin display alternative complementary expression patterns in nerve and muscle tissue [57], suggesting that combination of N-cadherin-dependent positive and T-cadherin-dependent negative navigation signals are necessary for correct wiring. In the chick optic tectum T-cadherin, N-cadherin and R-cadherin are expressed in distinct patterns during retinotectal synaptogenesis. Unlike the other two cadherins, T-cadherin is not concentrated at synapses and may play a role in limiting the arborization of retinal axons [60]. In the marmoset embryo cortical development appears to be coordinated by complex interplay between ten different cadherins including T-cadherin which are differentially expressed at specific stages and areas of the developing brain [61].

2.2. Neuropsychiatric Disorders

The potential physiological role of T-cadherin in the development of neural circuits is highlighted by data from genome-wide association studies (GWAS) aimed at identifying candidate susceptibility genes for neurobehavioral and neuropsychiatric disorders. Variations in *CDH13* gene have been linked to the risk of substance use disorders such as nicotine dependence [62-64], alcohol dependence [65-68], methamphetamine and other illegal substance dependence [69-74] and altered response to amphetamine [75, 76]. *CDH13* has been reported to be expressed in the groups of neurons implicated in drug response, reward system and cognitive modulation such as the substantia nigra pars compacta and ventral tegmentum [77]. Animal studies demonstrate that $Cdh13^{-/-}$ mice display evidence for reduced reward from a normally highly rewarding cocaine dose. This effect is accompanied by alterations in dopamine levels, dopaminergic fiber densities and expression of the activity dependent transcription factor npas4 that regulates the excitatory-inhibitory balance within neural circuits [78]. $Cdh13^{-/-}$ rats displayed altered reward-directed behaviors, including cue-induced reinstatement of cocaine seeking [79]. Together, these data suggest that T-cadherin may regulate the activity of the dopaminergic brain system and dopamine-associated behavior.

Abundant evidence points to the importance of T-cadherin in the pathogenesis of attention deficit hyperactivity disorder (ADHD), a complex childhood behavioral disorder with environmental and genetic etiology that is characterized by inappropriate levels of attention, hyperactivity, and impulsivity. Among suggested pathophysiological mechanisms contributing to ADHD are connectivity disturbances in fronto-striatal or meso-cortical neural circuits and dysfunctions in the mesolimbic dopaminergic system [80]. In 2008 *CDH13* was identified as one of the top genes associated with ADHD [81] and since then confirmed to be a potential risk gene for ADHD by many studies [82-87]. SNPs in *CDH13* gene were associated with alterations in neurocognitive functioning in ADHD patients [88], with risk tolerance and risk behaviors [89], and with extremely violent behavior in antisocial recidivistic offenders who frequently exhibit impaired impulse control and other ADHD symptoms [90]. *CDH13* variants have been also linked to autism spectrum disorder (ASD) which often shares traits with ADHD [91, 92], schizophrenia [93] and depression [66]. Mavroconstanti et al. performed sequencing of the *CDH13* gene in adult ADHD patients and healthy controls and attempted to characterize the identified *CDH13* variants by cloning and expression in CHO and HEK293 cells, but failed to detect any abnormalities in the processing or cellular localization of the mutant proteins (although effects on cellular functions such as neuronal migration have not been analysed) [94].

Significant progress has been made in understanding the mechanistic links between *CDH13* expression and the risk for neuropsychiatric disorders. Of particular interest are studies which utilize $Cdh13^{-/-}$ mice to study the role of T-cadherin in development of brain neural circuits. Analysis of the developing mouse brain demonstrates that in wild type (WT) mice *Cdh13* gene expression dynamically changes in the caudal-to-rostral direction and follows the trajectory pattern of serotonergic fibers. $Cdh13^{-/-}$ mice display increased density of 5-HT neurons in the dorsal raphe and higher serotonergic innervation of the prefrontal cortex. Since *Cdh13* gene is strongly expressed not only on 5-HT neurons but also on radial glial cells which play an important role in regulation of neuronal migration, these data suggest that T-cadherin regulates the development of the serotonergic system by participating in the spatiotemporal control of 5-HT neuron navigation [95]. Furthermore, T-cadherin is involved in the development of glutamatergic and GABAergic synapses [96] and negatively regulates hippocampal inhibitory GABAergic synapses, while its loss results in higher locomotor activity and alterations in learning and memory functions [97].

Interestingly, ADHD patients have been reported to have decreased serum APN levels, which are inversely correlated to psychiatric symptoms.

However, it is not clear whether adiponectin *per se* or its interactions with T-cadherin play any causative role in the pathogenesis of ADHD [98].

T-cadherin is also suggested to play a role in adaptation to stress and neuroprotection. $Cdh13^{-/-}$ mice exhibit decreased adaptive reactions caused by early-life stress leading to delayed habituation, no reduction of anxiety-like behaviour and decreased fear extinction [99]. T-cadherin deficiency was associated with altered endoplasmic reticulum function and expression of proteins involved in endoplasmic reticulum stress response including Grp78/BiP [99] which has previously been demonstrated to mediate T-cadherin signaling effects in the endothelium [100]. $Cdh13^{-/-}$ mice also exhibit an increased number of apoptotic cells and a concomitant decrease in amounts of interneurons and late-born pyramidal neurons in the cortex, thus pointing toward neuroprotective effects of T-cadherin during neuronal development [101]. Another study demonstrated that T-cadherin loss may contribute to cognitive and social behavioral deficits by affecting function of Golgi cells within the cerebellar cortex, which was evident from aberrant expression/localization of glutamate decarboxylase GAD67 involved in GABA synthesis and reduced spontaneous inhibitory postsynaptic current in Golgi cells in $Cdh13^{-/-}$ mice [102].

3. THE CARDIOVASCULAR SYSTEM

In vascular tissues T-cadherin is abundantly expressed on SMCs, ECs [43-46, 103, 104], pericytes and adventitial *vasa vasorum* [103]. It is not expressed on vascular adventitial fibroblasts *in vivo* [103]. In heart tissue T-cadherin is also expressed on cardiomyocytes [37, 51]. An increasing number of GWAS in human subjects have identified links between *CDH13* gene variants and blood pressure traits [105-108], blood lipid levels and composition [109, 110], coronary heart disease [111-114], ischemic stroke [115] and metabolic syndrome components [107, 112, 114-130]. Computational gene web analysis has also predicted that *CDH13* gene is vital to cardiovascular and metabolic diseases [131]. The effects of identified *CDH13* polymorphisms on expression, cellular distribution and biological functionality of mutant T-cadherin protein in the various T-cadherin expressing cardiovascular cell types remain completely unknown.

3.1. Endothelial Cells and Angiogenesis

The endothelium is a multifunctional endocrine organ strategically placed between the vessel wall and the circulating blood, and is able to respond to physical and chemical signals by production of a wide range of factors that regulate vascular tone, cellular adhesion, thromboresistance, SMC proliferation, and vessel wall inflammation [132]. The endothelium is the principal regulator of vascular homeostasis and acute and chronic endothelial stress is a key event in many vascular pathologies [133, 134]. Endothelial dysfunction, activation and damage associated with lipoprotein oxidation and inflammation importantly contribute to initiation and progression of atherosclerosis. Pathological angiogenesis, a morphogenetic process resulting from abnormal activation of ECs, is a hallmark of advanced atherosclerotic plaques predisposing them to rupture and thrombosis, and also a characteristic feature of cancer contributing to growth of solid tumors and metastasis [133, 134].

Upregulation of T-cadherin on ECs occurs *in vitro* under conditions of oxidative stress [135], and *in vivo* during atherosclerosis [103] and restenosis [136] as well as on intratumoral neovascular ECs in murine tumor models of breast cancer, melanoma, lung cancer and rhabdomyosarcoma [137, 138] and in human hepatocellular carcinomas [139-141]. A large body of evidence supports that T-cadherin is capable of regulating several EC functions relevant to vascular protection/repair and control of angiogenesis. T-cadherin overexpression in ECs promotes cell cycle progression and proliferation [142] and protects against apoptosis induced by oxidative or endoplasmic reticulum stress [40, 135, 143], with transcription factor thioredoxin-1 being an important determinant of redox-sensitive regulation of T-cadherin in ECs [143]. Homophilic ligation of T-cadherin molecules by recombinant T-cadherin protein or agonistic antibody induces a more polarized, motile phenotype and promotes cell migration [142, 144]. A consequence of T-cadherin ligation-dependent modulation of EC adhesive properties in a 2-dimensional monolayer culture system is induction of tubular structures and arrangement of ECs into multicellular interconnecting chains which form a capillary-like/networked pattern resembling the initial response of ECs to a pro-angiogenic environment [145]. Exposure of EC monolayer to preparations of endothelial-derived microparticles harboring T-cadherin on their surface (mimicking homophilic ligation) also induced EC arrangement into tubular angiogenic structures [146]. Use of 3D *in vitro* angiogenesis models (EC-spheroid and heart tissue) demonstrated increased active outgrowth of newly formed capillary-like sprouts following either upregulation of T-cadherin on ECs or homophilic ligation through embedment of spheroids/tissue in substrata containing recombinant T-cadherin protein [145].

Relevant *in vivo* functions for T-cadherin in angiogenesis/revascularization have been demonstrated in murine models commonly used for investigating tumor development (the mouse mammary tumor virus (MMTV)-polyoma virus middle (pY-V-mT) transgenic model) (MMTV-PyV-mT) [35], peripheral artery disease (hind limb ischemia model) [50], cardiac hypertrophy (chronic pressure overload induced by transverse aortic constriction (TAC)) [51], and skeletal neovascularization (employing myoblast-mediated gene transfer into mouse skeletal muscle) [145]. In the mammary tumor model, T-cadherin gene ablation restricts tumor vascularization and expansion and also limits hypoxia-induced retinal angiogenesis [35]. In the hind limb ischemia model $Cdh13^{-/-}$ mice displayed impaired revascularization/blood flow recovery [50]. In the model of cardiac hypertrophy $Cdh13^{-/-}$ mice displayed reduced capillary density after chronic pressure overload [51]. Implantation of myoblast clones co-expressing VEGF and secreted T-cadherin protein induces formation of capillaries of larger diameters (without affecting vessel density) compared with respective clones expressing the same dose of VEGF alone. Importantly, clones expressing only T-cadherin protein affected neither density nor caliber of neovessels, suggesting that *in vivo* T-cadherin is not a primary angiogenic stimulus, but rather a modulator of angiogenesis that requires initial destabilization of the vessel by growth factors (e.g., VEGF) in order to exert its actions on EC phenotype conversion, migration, proliferation and survival [145].

Intracellular signal effectors participating in T-cadherin-ligation-dependent modulation of EC angiogenic behaviors have been identified through *in vitro* studies, and include PI3K/Akt/mTOR/GSK3β pathway [41, 135], β-catenin [41], small GTPases RhoA and Rac1 [144], p38MAPK [135], and the protein kinase RNA-like endoplasmic reticulum kinase (PERK) branch of the unfolded protein response [147]. Domains EC1 and EC5 of T-cadherin were found essential for its proangiogenic effects [148]. Membrane proximal molecules with which T-cadherin can directly associate in order to initiate transmembrane signal transduction in ECs following T-cadherin engagement include stress chaperone GRP78/BiP, integrin-linked kinase and integrin β3 [40, 41]. An alternate model put forward for the proangiogenic properties of T-cadherin relates to the function of T-cadherin as a receptor for APN [35, 47, 50, 51, 149], the ability of APN to sequester a variety of growth factors/angiogenic cytokines [150], the essential requirement of T-cadherin for APN-mediated enhancement of angiogenic differentiation of ECs *in vitro* [50, 151] and for APN-mediated revascularization [50] and cardioprotection [51] *in vivo*. It has been hypothesized that T-cadherin-APN binding interaction on ECs enables T-cadherin to transmit transmembrane angiogenic signals through growth factors/angiogenic cytokines that are complexed with APN [35] or that APN signals through membrane proximal signaling molecules associated with T-cadherin (e.g., integrin linked kinase) [51].

Mechanisms whereby T-cadherin might co-opt or cross-talk with APN-associated factors to regulate angiogenic behavior are far from clear. APN has been shown to possess proangiogenic functions *in vitro* and *in vivo* and to promote activation of AMPK, Akt and eNOS [151]. APN-induced eNOS activation and functional responses to APN *in vitro* were abolished by transduction with either dominant-negative AMPK or dominant-negative Akt. Dominant-negative AMPK also inhibited APN-induced Akt activation, and PI3K inhibition or dominant-negative Akt inhibited eNOS activation and functional responses to APN without affecting AMPK phosphorylation [151]. This study suggests that AMPK is upstream of Akt and that APN-induces angiogenesis by promoting cross-talk between AMP-activated protein kinase and Akt signaling within ECs. Although involvement of T-cadherin was not investigated in this study, the use of T-cadherin null mice demonstrated that APN requires T-cadherin to stimulate AMPK signaling [51]. Further, siRNA-mediated T-cadherin knock-down in ECs inhibits the ability of APN to promote migration and mitosis [50]. Since Akt axis signaling is a major transduction node activated in response to homophilic T-cadherin ligation in ECs [40, 41, 135] and homophilic ligation induces angiogenic responses even in the absence of APN [42, 145, 146], it is conceivable that APN might "piggy back" onto T-cadherin-ligation mediated PI3K/Akt/mTOR/GSK3β pathway activation. Whatever the mechanism, it can be assumed that heterophilic T-cadherin-APN interactions and homophilic T-cadherin-T-cadherin interactions and effects on intracellular signaling and cell function are not mutually exclusive; their independent functionality or interdependent cross-talk may depend on their relative expression levels under any given physiological or pathophysiological condition.

Another aspect of the role for T-cadherin in endothelial (patho)biology concerns its presence in the circulation as a component of EC-derived microparticles (MPs). MPs are small submicron plasma membrane-derived vesicles which originate from many cell types through exocytotic budding and shedding from cell membranes into blood and body fluids under physiological and pathological conditions [152-154]. They harbor biological information such as proteins (signal proteins and receptors, cytoskeleton and effector proteins), lipids, and nucleic acids, (e.g., microRNA, mRNA, DNA fragments) which can be transferred to proximal or remote cells (homologous or heterologous) without direct cell-to-cell contact. In addition to their biological actions in inflammation, immune responses and coagulation MPs are capable of directly stimulating intracellular signaling and eliciting cellular responses, such as proliferation, survival, adhesion, chemotaxis, and intercellular communication [152-154]. Endothelial-derived MPs are markers of endothelial dysfunction, and are considered to play a major biological role in inflammation, vascular injury, angiogenesis, and thrombosis [155, 156]. *In vitro* studies have demonstrated that T-cadherin protein is present on MPs released from stressed/apoptotic ECs (exposed to TNFα, H_2O_2, or thapsigargin); these MPs can induce homophilic-ligation T-cadherin-dependent signaling (e.g., activation of Akt axis signaling) and a proangiogenic functional response in target ECs. In a study on young healthy individuals and a patient cohort characterized with respect to the stage of atherosclerosis and presence of endothelial dysfunction assessed by reactive hyperemia peripheral arterial tonometry, the presence of T-cadherin in plasma as a component of MPs was associated with endothelial dysfunction [146]. The level of T-cadherin harboring MPs was increased in patients with early atherosclerosis and with chronic coronary artery disease, and in these groups Spearman's correlation also revealed a significant dependency of T-cadherin release into the circulation and the degree of endothelial dysfunction. The presence of T-cadherin in plasma of patients with early atherosclerosis supports that upregulation and shedding of T-cadherin from ECs is an initial response of the endothelium to activation and stress.

3.2. Smooth Muscle Cells and Atherosclerosis

SMCs are the most abundant cell type in the blood vessel wall, and their primary function is to maintain vascular homeostasis through coordinated cycles of contraction and relaxation. Unlike the majority of cells in the adult organism which are terminally differentiated, vascular SMCs retain inherent plasticity even in the mature vessel and can undergo reversible changes in phenotype in response to changes in local environmental cues [157-161]. The differentiated SMC phenotype is characterized by expression of a unique repertoire of contractile proteins, myofilaments and signaling molecules necessary for regulation of smooth muscle contractility. The dedifferentiated SMC phenotype is characterized by decay of contractile markers and acquisition of synthetic, migratory and proliferative properties required for vascular development, reparation after vascular injury and remodeling in response to altered blood flow. Deregulated SMC phenotype switching and failure to maintain/regain the differentiated state is a pathogenic basis for development of several vascular disorders such as atherosclerosis, post-angioplasty restenosis and hypertension.

Associations between altered expression of T-cadherin on SMCs and pathogenesis of occlusive vascular disorders have been reported in a number of studies. *CDH13* gene expression was found to be lower in porcine atherosclerosis-prone coronary arteries than in atherosclerosis-resistant mammary arteries [162], suggesting T-cadherin expression negatively associates with predisposition to atherosclerosis. However, several immunohistochemical and biochemical investigations rather support a positive association between T-cadherin expression and occlusive vascular disease. Increased T-cadherin expression was observed on intimal SMCs in the model of experimental restenosis during the early phase of reparative proliferation [136], in human aortic and coronary atherosclerotic lesions at different stages of disease progression [103, 104, 163], and also in cardiac allograft vasculopathy (with restenosis-like remodeling) and ischemic cardiomyopathy (with atherosclerosis-like remodeling) in a rat allograft model [164]. T-cadherin protein expression was also increased in aortic tissue from atherosclerotic apolipoprotein E (*ApoE*)-knockout mice as compared to that in WT mice [53]. Upregulation of T-cadherin transcript and protein expression in SMCs *in vitro* under different conditions of *in vivo* pathological stress (e.g., hyperglycemia, hyperinsulinemia, oxidative stress) [165] further supports elevation of T-cadherin on SMCs as a molecular component of occlusive vascular disease.

A role for T-cadherin in control of phenotype was initially suggested following immunohistological observations of upregulation of T-cadherin protein on SMCs in lesional arterial tissues from human atherosclerosis and experimental restenosis [103, 136], prominent inverse staining intensities for T-cadherin and SMC contractile marker alpha–smooth muscle actin (α-SMA) in diseased human aorta [103], and positive association between staining intensities for T-cadherin and proliferating cell nuclear antigen [103, 136]. Subsequent gain- and loss-of function studies *in vitro*, variously using rat, porcine, murine and human aortic SMCs, provided direct evidence that T-cadherin is a molecular determinant of SMC phenotype. Overexpression of T-cadherin in SMCs prompts phenotype shifting toward dedifferentiation as manifested by loss of the typical differentiated "spindle" morphology, reduced filamentous actin/stress fibers and decay of differentiated SMC contractile markers (α-SMA, *sm*MHC and *h*-caldesmon) [166]. Further, T-cadherin upregulation altered abundance/cellular distribution of myocardin and myocardin related transcription factors [166], which are transcriptional coactivators of the serum response factor-dependent transcriptional regulatory program that critically controls SMC differentiation status [167-169]. Importantly, T-cadherin gene ablation *in vitro* enforced the morphological and contractile molecular signature of the differentiated phenotype [166], while a reduction in SMC T-cadherin expression accompanied transforming growth factor β1-induction of the contractile phenotype [53]. These findings indicate that although T-cadherin is ubiquitously expressed by SMCs in healthy vessels it is not essential for mature SMC contractile function but may rather play a permissive role with respect to control of direction and extent of SMC phenotype switching.

T-cadherin upregulation on SMCs *in vitro* is accompanied by several behavioral alterations consistent with SMC modulation to the differentiated phenotype and adoption of reparative/remodeling functions. These include the following: reduced intrinsic contractile competence [165]; increased rates of cell cycle progression and proliferation [166, 170]; reduced ability to adhere and spread on extracellular matrix substrata in association with an increased capacity for motility and directional/polarized migration [166, 171]; gain of matrix remodeling capacity manifested by collagen fibril reorganization in 3D-SMC spheroid cultures [165] and increased expression of matrix metalloproteinase 2 in monolayer SMC cultures [166]. Upregulation of T-cadherin also increases pro-survival autophagy [172], which has been proposed to serve as a major proteolytic mechanism to remove contractile proteins during phenotype transition to dedifferentiation and to be important to the development or maintenance of the dedifferentiated phenotype [173, 174]. Molecular mechanisms and signal transduction pathways involved in mediating the many functions of T-cadherin in SMCs have been extensively reviewed [175, 176] and will only be outlined herein. PI3K/Akt is a major signal transduction node utilized by T-cadherin, with activation of GSK3β and mTORC1/S6K1 downstream effector branches [165, 166]. GSK3β inactivation is the dominant mediator of T-cadherin-induced dedifferentiation and proliferation [166]. Pro-survival autophagy depends on activation of MEK1/2/Erk1/2 pathway signaling [172]. Phenotype-associated functional transition from static cell anchorage (differentiated) to migration (dedifferentiated) involves adhesion molecule expression, cell-matrix adhesion, organization of intercellular contacts and control of intracellular tension forces generated by the actin cytoskeleton [177].

An unclear issue regarding the role of T-cadherin in SMCs is its putative function as a receptor for LDL, an established risk factor for atherosclerotic disease [178, 179]. Although LDL was identified as a heterophilic ligand for T-cadherin expressed in aortic medial tissue (i.e., predominantly composed of SMCs) almost 25 years ago [44-46], the physiological relevance of heterophilic LDL-T-cadherin interaction to vascular tissue remains very poorly defined. Ectopic overexpression of T-cadherin in HEK293, EA.hy926 and cadherin deficient L929 fibroblast cell lines enhanced LDL-induced signaling (intracellular Ca^{2+} mobilization, Erk1/2 MAPK and nuclear translocation of nuclear factor κβ) and functional responses (proliferation and migration) to LDL [180-182]. In SMCs LDL is capable of activating a number of intracellular processes variously linked to its ability to stimulate vasoconstriction or proliferation (e.g., intracellular Ca^{2+} mobilization, cytosolic alkalinzation, phosphoinositide turnover, activation of protein kinase C, S6-kinase and Erk1/2 MAPK, generation of reactive oxygen species, DNA synthesis, nuclear proto-oncogene expression, *inter alia*) [39, 183-188]. Establishing whether these responses of SMCs to LDL indeed involve direct T-cadherin-LDL binding interactions would require appropriate gain- and loss-of function in this cell type. Nevertheless, by analogy to the aforementioned data obtained through ectopic upregulation of T-cadherin in other cell lines, one might speculate that T-cadherin-LDL binding can facilitate LDL-dependent vasoconstriction and mitogenic signaling.

3.3. Vascular Insulin Resistance

Expression of T-cadherin on ECs and SMCs is elevated *in vitro* under conditions of stress (e.g., oxidative, hyperglycemic, hyperinsulinemic) [135, 165], and *in vivo* in vascular disorders such as atherosclerosis [103] and restenosis [136] that are associated with insulin resistance. A relationship between T-cadherin expression on vascular cells and development of insulin resistance has been demonstrated, whereby T-cadherin upregulation or silencing respectively attenuated or enhanced insulin responsiveness. The effects of T-cadherin ligation on insulin sensitivity in vascular cells depend upon its ability to directly associate with insulin receptor within lipid rafts, and its utilization of PI3K/Akt/mTORC1 axis signaling and downstream targeting of components (e.g., S6K1, eNOS) common to insulin signaling [42]. Increased cell surface T-cadherin expression caused by oxidative stress, inflammation, or prolonged exposure to insulin results in chronic stimulation of the Akt cascade, which in its turn induces compensatory hyperactivation of the negative feedback loop of the insulin cascade leading to increased S6K1-mediated serine phosphorylation and degradation of insulin receptor substrate 1 (IRS-1), and subsequent attenuation of insulin signaling through IRS-1/PI3K/Akt [42, 165]. Insulin resistance in T-cadherin overexpressing ECs is functionally manifest as an attenuation of insulin-dependent eNOS phosphorylation and activation (i.e., reduced NO bioavailability), cell migration and angiogenesis [42]. These processes normally promote vascular quiescence and healing/ endothelial regeneration [189]. In T-cadherin overexpressing SMCs insulin resistance functionally results in a reduction in insulin-responsive contractile competence [165]. Insulin-induced IRS-1/PI3K/Akt axis signaling in SMCs is important to maintenance of the differentiated, contractile phenotype [190, 191]. Thus, although T-cadherin-dependent activation of pro-survival/reparative pathways in vascular cells during initial stress or injury is advantageous to preserved vascular function [135, 172], in a setting of sustained stress chronic ligation of T-cadherin would compromise insulin-mediated vasculoprotective functions and exacerbate disease progression. $Cdh13^{-/-}$ mice have normal peripheral insulin sensitivity *in vivo*, based on hyperinsulinemic-euglycemic clamping (which measures glucose uptake) [192]. $Cdh13^{-/-}$ mice have not yet been examined for vasomotor sensitivity to insulin *in vivo*. However, based on *in vitro* observations of elevated levels of IRS-1 and enhanced insulin-dependent signaling in T-cadherin-silenced ECs and SMCs [42, 165] one may assume that insulin receptor/IRS-1 levels coupling and thus insulin sensitivity *in vivo* are at least intact.

Another effect of T-cadherin on vascular NO bioavailability, albeit independent of insulin, has been demonstrated in a study using *ex vivo* aortic ring segments from WT and $Cdh13^{-/-}$ mice [193]. T-cadherin gene ablation was associated with a reduction in basal tissue accumulation of NO and impaired vasorelaxation induced by acetylcholine (which increases endogenous NO). Aortic tissue homogenates from WT and $Cdh13^{-/-}$ mice exhibited comparable levels of phospho-eNOS^{Ser1177} indicating comparable eNOS activity. Reduced NO bioavailability was due to superoxide-mediated inactivation of NO and increased EC caspase 3 activity/apoptosis arising through suppression of Akt phosphorylation/activation. The impact of T-cadherin deficiency on Akt activity (reduced) and apoptosis (increased) in aortic tissue is consistent with *in vitro* studies demonstrating that Akt is a major signal target downstream of T-cadherin in vascular cells and that pro-survival properties of T-cadherin in ECs are mediated *via* Akt/mTOR and Akt/GSK3β signal pathways [41, 42, 135, 166].

3.4. Cardiomyocytes

T-cadherin is expressed on C2C12 myotubes *in vitro* and on cardiomyocytes *in vivo*, where it exhibits a globally punctuate distribution (in keeping with its location to lipid raft domains in the plasma membrane

[37]), and colocalizes with APN [51]. The physiological function of T-cadherin in cardiac tissue was investigated by examining myocardial responses of $Cdh13^{-/-}$ mice after TAC (which induces cardiac hypertrophy and heart failure) and after ischemia-reperfusion injury (which induces myocardial infarction) [51]. Myocardial responses to both short-term TAC (compensated hypertrophy, manifest as hypertrophy in association with normal contractility) and long-term TAC (decompensated hypertrophy, manifest as impaired contractility, increased ventricular dilation, reduced neovascularization and cardiomyocyte apoptosis) were exaggerated in $Cdh13^{-/-}$ mice. The response to ischemia-reperfusion injury, evaluated as myocardial infarct area and extent of apoptosis, was also exaggerated in the $Cdh13^{-/-}$ mice. The structural and functional responses of APN knockout ($AdipoQ^{-/-}$) mice to TAC and ischemia-reperfusion injury mice mirrored those of $Cdh13^{-/-}$ mice. Adenoviral administration of APN prior to cardiac stress prevented exaggerated hypertrophic and ischemia-reperfusion phenotypes and the associated reduction in $AdipoQ^{-/-}$ mice, but not in $Cdh13^{-/-}$ mice or in $AdipoQ^{-/-}$- $Cdh13^{-/-}$ double knockout mice. The presence of T-cadherin was also necessary for APN-dependent stimulation of AMPK [51]. In rats, volume overload induced by infrarenal aorta-vena cava fistula resulted in reductions in serum and myocardial APN levels, myocardial APN receptor (AdipoR1/R2 and T-cadherin) levels, and myocardial AMPK activity; dysfunction of ventricular myocytes isolated 12-weeks post fistula was demonstrated [194]. Thus, T-cadherin in conjunction with APN protects the heart from multiple stressors, such as chronically increased afterload and myocardial ischemia. How T-cadherin actually signals its cardioprotective contribution following sequestration of APN has not been delineated.

4. LUNG

T-cadherin is expressed on pulmonary epithelial cells and pulmonary ECs [195]. Genetic variants of *CDH13* gene have been reported to determine Chinese individuals' susceptibility to chronic obstructive pulmonary disease (COPD) [196]. Short-acting β2-agonist bronchodilators are the most common medications used in COPD, and based on the quantitative spirometric response to inhaled β2-agonists, genetic variants of *CDH13* determining bronchiodilator responsiveness have been identified [197]. An investigation on susceptibility to air pollution found that effects of particulate matter with an aerodynamic diameter ≤10μm on lung function decline was modified by *CDH13* genetic variants in Korean men [198]. T-cadherin have been demonstrated to downregulate surfactant protein D production in A549 lung cancer cell line with alveolar type-II cell characteristics *in vitro* [199]; however, physiological relevance of this phenomenon in bronchioloalveolar cells remained unknown.

Investigations on the functional role of T-cadherin in the (patho)physiology of lung function are limited and restricted to evaluating the relevance of T-cadherin as a receptor for APN. APN exerts beneficial effects on allergic airways responses [200], and available data suggests that T-cadherin may be important for APN transport into the lungs and for mediating some of the beneficial effects of APN on allergic airways responses. APN was lower in the bronchoalveolar lavage (BAL) fluid of naïve *Cdh13*$^{-/-}$ mice compared with that of WT mice suggesting a role for T-cadherin in transporting APN across the alveolar capillary barrier through a vesicular transcytosis pathway involving binding of APN to T-cadherin expressed on ECs [201]. However, after allergen (ovalbumin) or ozone exposure BAL APN concentrations were higher in *Cdh13*$^{-/-}$ mice [201, 202], indicating that, at least in the setting of lung injury/increased capillary permeability, diffusion/leakage of APN through paracellular pathways between ECs into the alveolar spaces rather than T-cadherin-mediated vesicular transport dominates movement of APN from the blood into the lungs. Compared with WT animals, *Cdh13*$^{-/-}$ mice exhibited reduced allergic airways disease inflammatory responses to ovalbumin sensitization and challenge as evidenced by a marked reduction in allergen induced airway hyperresponsiveness, eosinophil recruitment to the airways, Th2 cytokine expression, mucous cell hyperplasia and BAL IL-17A expression [201]. *AdipoQ* $^{-/-}$- *Cdh13*$^{-/-}$ double knockout mice reversed the effects of T-cadherin deficiency alone, indicating that T-cadherin did not mediate the ability of APN to reduce allergic airways responses and that the protective effects of T-cadherin deficiency required APN [201]. It was suggested that effects of T-cadherin deficiency likely reflects activation of other APN signaling receptors/pathways (such as AdipoR1 or AdipoR2) secondary to elevated serum APN concentrations observed in $Cdh13^{-/-}$ mice [35, 50, 51, 201, 202]. In contrast to allergen challenge, APN binding to T-cadherin was found to be required for APN-suppression of ozone-induced inflammation: subacute ozone exposure inflammatory responses (IL-17A mRNA expression, terminal bronchiolar lesions, and *saa3* expression) were increased in $Cdh13^{-/-}$ mice and in $AdipoQ^{-/-}$ mice (vs. WT), but were not further increased in the $AdipoQ^{-/-}$- $Cdh13^{-/-}$ double knockout mice [202].

5. PANCREAS

The islets of Langerhans are the functional units of the endocrine pancreas and have a paramount role in maintaining glucose homeostasis through regulation of insulin secretion. An important function for T-cadherin in regulating pancreatic insulin secretion has been identified by Tyrberg and colleagues [192]. In the pancreas T-cadherin localizes not only to the exocrine vasculature, but also to endocrine islet β-cells where it colocalizes with dense-core insulin-containing granules. *In vitro* comparison of the insulin-release properties of pancreatic islets isolated from WT and $Cdh13^{-/-}$ mice revealed that islets lacking T-cadherin expression exhibit a normal first rapid phase insulin release response whereas the second prolonged phase insulin release response was impaired, resulting in a 58% reduction of insulin secretion. *In vivo*, $Cdh13^{-/-}$ mice exhibit severely impaired or delayed second phase insulin secretion during hyperglycemic clamp. $Cdh13^{-/-}$ mice show normal islet architecture and insulin content, and a normal insulin sensitivity and glucose utilization in peripheral tissues, such as muscle. However, intraperitoneal glucose tolerance testing demonstrated that $Cdh13^{-/-}$ mice develop glucose intolerance, which may arise through inadequate insulin secretion [192]. In addition to demonstrating the requirement of T-cadherin for sufficient and persistent glucose-stimulated insulin release by pancreatic β-cells, this study introduces a potentially novel function for T-cadherin in vesicular trafficking and exocytosis.

6. LIVER

In normal liver samples, T-cadherin is expressed in ECs of large blood vessels and in myofibroblasts, weakly expressed in sinusoidal ECs, and absent in hepatocytes [140]. Hypomethylation of *CDH13* gene has been linked to liver cirrhosis [203]. Functional genomic screening identified T-cadherin as a molecule upregulated in liver-function-inducing stromal cells [204]. A role for T-cadherin in modulating hepatic functions has been explored [205]. Primary hepatocytes were cocultured with mock- or T-cadherin-transduced CHO cells on a substratum containing immobilized recombinant T-cadherin. Cellular or substratum presentation of T-cadherin to hepatocytes enhanced liver-specific functions as assessed by measuring albumin secretion, urea synthesis, and cytochrome P450 1A1 activity as surrogate markers for liver-specific protein synthesis, nitrogen metabolism, and detoxification functions. There are no further studies elucidating the role for T-cadherin in regulating hepatic phenotype.

7. KIDNEY

T-cadherin displays a distinct and dynamic expression pattern during the differentiation of foetal human glomeruli. At the early capillary loop stage it is expressed apically on visceral epithelial cells of Bowman's capsule which start to differentiate into podocytes, while at the advanced capillary loop stage it is restricted to the foot processes of the podocytes within the glomerular filtration barrier, suggesting a role in podocyte differentiation and the formation of the glomerular capillary network [206].

8. BONE AND CARTILAGE

Microarray analysis has identified T-cadherin as a gene that might play a role in bone development and regeneration [207]. Downregulation of T-cadherin mRNA and protein was observed in chondrocytes from patients with knee osteoarthritis and suggested to contribute to the loss of protective effects of adiponectin in the osteoarthritic tissue [208].

9. THE COCHLEA

In the cochlea of postnatal and adult mice T-cadherin mRNA and protein were expressed in the organ of Corti, the spiral ganglion, and the stria vascularis. The protein was located apically on the inner and outer hair cells [209]. In rat cochlea T-cadherin was present in fibrocytes or pillar cells; its expression pattern was mutually exclusive with E- and N-cadherin and dynamically changed during embryonic development suggesting that it may regulate the emergence of specific cell phenotypes and cell differentiation [210].

10. CANCER

Cadherins are well recognized as key determinants of cancer progression. Epithelial-to-mesenchymal transition (EMT) characterized by downregulation of epithelial markers such as the prototype cadherin family member E-cadherin, with simultaneous upregulation of mesenchymal markers including N-cadherin, leads to malignant transformation and acquisition of invasive properties by carcinoma cells. *CDH13* gene is mapped to the chromosome region 16q24 which displays frequent loss of heterozygosity (LOH) in cancer suggesting a potential role for T-cadherin as a tumor suppressor. Indeed, inactivation of *CDH13* due to allelic loss or hypermethylation of the promoter region is frequently observed in solid tumors. However, accumulating data from experimental studies suggest that T-cadherin-dependent regulation of cancer progression is a complex process which depends on the tumor type and involves both control of tumor cell function and cross-talk with the tissue microenvironment. Below we discuss the available data for T-cadherin expression and function in different types of cancer.

10.1. Cutaneous Cancer and Other Proliferative Conditions of the Skin

The epidermis, the most outward layer of the skin which protects the body from environmental factors, is a very dynamic tissue that undergoes constant turnover. Cell renewal in the epidermis is initiated by keratinocytes residing in the lower *stratum basale* layer which are mitotically active and possess stem cell-like properties. Daughter cells generated by basal keratinocytes detach from the basal membrane, move upwards, differentiate and create novel cell strata thus ensuring tissue renewal and regeneration. Keratinocytes are the predominant cell population in the epidermis, with other types including melanocytes, Merkel-Ranvier cells and Langerhans cells [211].

The relationship between T-cadherin expression and proliferative/malignant disorders of the epidermis is a good example of complex and differential effects exerted by this protein on different cell types, even in the same tissue. The data from genetic association and immunohistochemical studies display significant discrepancies which might be attributed to heterogeneity of the analysed samples, cell and tumor types (as well as, possibly, to methodological issues such as the use of different antibodies).

In the healthy skin T-cadherin expression is strongest in the basal keratinocyte layer and gradually diminishes as new cells move upward to the skin surface and differentiate [212-214]. Since the cells in the basal layer exhibit the highest proliferation rates compared to their more differentiated descendants, this is the first indication that T-cadherin does not necessarily act as a suppressor of keratinocyte proliferation. T-cadherin is also present in the basal layer of sebaceous glands, on myoepithelial cells of apocrine glands, in the secretory coils and excretory ducts of eccrine glands, on melanocytes [215], and in the skin stem cell niches of the hair follicle [214, 216].

In basal cell carcinoma (BCC), the most common type of skin cancer which originates from basal keratinocytes, LOH in intron 1 of the T-cadherin gene or aberrant methylation of the T-cadherin promoter region was found in 20% and 24% cases, respectively [217]. Decreased protein expression in tumor samples was also observed; however, another study reported the opposite, namely, strong upregulation of T-cadherin protein in BCC irrespective of the histological type including superficial, nodular and infiltrative morphology, with highest staining intensities at intercellular borders at the leading invasive fronts of the tumors [212]. While BCC progresses slowly and rarely metastasizes, squamous cell carcinoma (SCC) often shows aggressive behaviour with high metastatic potential. In contrast to BCC, SCC is characterized by hypermethylation of *CDH13* gene [218] and loss or aberrant expression of T-cadherin protein which correlates with histological features of hyperproliferative, poorly differentiated and invasive tumors [219, 220]. Common precursors for SCC include actinic keratosis and Bowen's disease which are considered low-risk and high-risk lesions for malignant transformation, respectively. T-cadherin protein is mostly retained in actinic keratosis although some regions displaying loss of expression in the basal layer of the epidermis have been shown to give rise to SCC tumors [219, 221]. In Bowen disease a markedly weakened expression of T-cadherin on the basal cell layer was observed [221]. Taken together, these data demonstrate T-cadherin downregulation plays a role in malignant transformation and progression of SCC but not BCC.

Experimental studies demonstrate that T-cadherin loss promotes proliferation, migration and invasion of keratinocyte and SCC cell lines *in vitro* [219, 222], as well as tumor growth and metastasis *in vivo* [223, 224]. Inhibitory effects of T-cadherin on SCC cells are likely to be mediated by regulation of epidermal growth factor receptor (EGFR) phosphorylation and membrane compartmentalization, as well as modulation of cell-matrix adhesion through control of surface levels of β1 integrin [225-227]. Paradoxically, T-cadherin overexpression also stimulated growth of SCC tumors *in vivo*, which occurred *via* enhanced intra-tumoral angio/lymphangiogenesis associated with increased VEGF expression by T-cadherin-overexpressing SCC cells [223]. This dual effect of T-cadherin on cancer progression was confirmed in studies on breast cancer (below).

T-cadherin expression is frequently lost in melanoma [215]. Downregulation of T-cadherin in melanocytes and melanoma cells promoted invasion, while re-expression inhibited anchorage-independent growth, migration, invasion, and tumor xenograft growth [215, 228, 229]. Inhibition of tumor growth by T-cadherin *in vivo* has been attributed to stimulation of apoptosis *via* attenuation of AKT/CREB/AP-1/FoxO3a signaling pathway [229]. *CDH13* gene expression in melanoma cells is under the control of transcription factor BRN2 which might participate in malignant transformation of melanoma by repressing T-cadherin levels in melanocytes [230]. Another study, however, demonstrated that T-cadherin may promote melanoma progression by recruiting stromal cell components to tumors [231].

T-cadherin expression in the basal keratinocyte layer is also decreased in psoriatic skin concomitantly with upregulation of P-cadherin which is considered to be a reflection of keratinocyte hyperproliferation in psoriasis vulgaris [232].

10.2. Breast Cancer

In the healthy breast tissue T-cadherin is expressed on the myoepithelium and on the mammary ductal epithelial cells where it displays a location on the apical surface of polarized cells characteristic for GPI-anchored proteins but not for classical cadherins [35]. Loss of T-cadherin expression attributed to aberrant *CDH13* promoter methylation, rather than to the presence of mutations in *CDH13* gene, was observed in human breast cancer cell lines, primary tumor tissue and metastases [233-

241]. The absence of T-cadherin expression occurred early during transition from premalignant lesions to invasive carcinoma [242], correlated with increased tumor size, lymph-vascular invasion, higher disease stage and poor prognosis in patients with axillary lymph node-positive breast cancer [243], with tumor grade and negative prognosis in triple-negative breast cancer [235, 244], with response to neoadjuvant chemotherapy for locally advanced breast cancer [245], and overall survival [246]. *CDH13* methylation was more prevalent in HER2-positive tumors [247, 248], correlated with estrogen receptor (ER) expression, and poor differentiation in ER-negative tumors was associated with decreased levels of *CDH13* gene expression [249]. The anti-tumor effects of cerivastatin in breast cancer cells were demonstrated to be accompanied by upregulation of T-cadherin expression [250].

Transduction of breast cancer cells with T-cadherin cDNA inhibited tumor growth and invasion *in vitro* and *in vivo* [233, 251]. Subsequent studies however demonstrated that, similarly to cutaneous carcinomas, the relationship between T-cadherin expression and breast cancer progression is complex and non-linear. The study of Hebbard et al. which utilized the MMTV-PyV-mT transgenic mouse model [35] demonstrated that T-cadherin expression was lost from developing mammary gland tumor tissue but was retained on ECs. Crossing the MMTV-PyV-mT model with $Cdh13^{-/-}$ mice demonstrated that T-cadherin deficiency reduced tumor growth by limiting tumor vascularization. However, at the same time it resulted in the appearance of a poorly differentiated tumor phenotype which actively metastasized to the lungs.

10.3. Prostate Cancer

CDH13 gene expression was reported to be decreased in prostate cancer (PCa) [252, 253]. Conversely, another study reported methylation of T-cadherin gene in only 31% of prostate cancers and in 20% of benign prostate disease [254]. Immunohistochemical analysis demonstrated that in contrast to gene expression levels which did not significantly correlate with tumor progression, T-cadherin protein expression was drastically increased in early stages of cancer. Its expression was more prominent in organ-confined than in advanced hormone-resistant tumors, correlated negatively with the Gleason pattern and reflected luminal/basal differentiation cell status [255]. T-cadherin overexpression induced the loss of epithelial polarity in prostate cell line BPH-1 and metastatic PCa cell line DU145, promotes proliferation, invasion, 3D organoid growth and transmigration through EC monolayers by regulating activity of EGFR and insulin growth factor-1 receptor, and modulates PCa cell response to doxorubicin [255, 256]. In the prostate tissue *CDH13* gene expression is under control of androgens [252] and estrogen receptor ERβ [257] and contributes to acquisition of a basal stem cell gene signature [258, 259]. Together, these data suggest that T-cadherin expression dynamically changes during PCa progression and may differentially influence cancer cell behavior depending on the stage, differentiation and hormonal status of the tumor. Promoter methylation of *CDH13* has also been reported in the serum of patients with PCa and was associated with clinicopathological tumor features and prognosis [260, 261].

10.4. Lung Cancer

CDH13 gene was found to be often inactivated in lung cancer specimens and cell lines due to both promoter methylation and deletion of locus [262]. Most studies have demonstrated a relationship between the loss of *CDH13* gene expression and non-small cell lung cancer (NSCLC) [263-278, 279] where *CDH13* promoter methylation occurred more frequently compared with small cell lung cancers [234], reflected responsiveness of NSCLC patients to chemotherapy [280], and correlated with levels of folate [281] which has been implicated in chemotherapy response and the homeostasis of DNA methylation. Allelic frequencies of several SNPs located in *CDH13* gene promoter and intron regions were significantly associated with different pathologic stages of NSCLC [282]. *CDH13* gene was more hypermethylated in poorly differentiated NSCLC than in moderately and highly differentiated tumors [283], in EGFR WT tumors as compared to those expressing EGFR mutations [284], as well as in adenocarcinoma as compared to SCC [285], correlated with invasiveness of lung adenocarcinomas [271], and was observed not only in tumors but also in plasma of NSCLC patients [286]. Data for relationships between loss of *CDH13* gene expression and overall survival of lung cancer patients are variable and have demonstrated correlation between poor overall survival and hypermethylation of *CDH13* gene [283, 284, 287, 288], between poor overall survival and hypermethylation of *CDH13* gene only in combination with *CDH1* gene hypermethylation [263], or no correlation between *CDH13* gene methylation status and survival [268, 270]. *Cdh13* gene hypermethylation has also been reported in animal models of spontaneous or carcinogen- and inflammation-induced lung tumors [289, 290] and facilitated local growth of transplanted NSCLC tumors in mice [265]. Frequent *CDH13* gene promoter methylation could be detected also in bronchial lavage of NSCLC patients [291]. Taken together, these data suggest that the loss of *CDH13* gene expression may serve as a diagnostic marker to detect early NSCLC.

10.5. Gastrointestinal Cancers

CDH13 gene methylation was observed in early Barrett's esophagus-associated neoplastic progression and in esophageal cancer [292-294], correlated with the stage and degree of tumor differentiation [295, 296], and clearly demarcated esophageal adenocarcinoma from esophageal SCC and normal esophagus [294].

LOH which includes the locus for the *CDH13* gene was demonstrated in gastric cancer tissues and was associated with lymphangial invasion of the tumors. Gastric cancer cell lines indeed displayed decreased levels of T-cadherin mRNA, although no mutations or abnormalities in the methylation status of the promoter region of *CDH13* gene were detected [297]. Another study, however, demonstrated abnormal methylation of the *CDH13* gene promoter in 35% gastric cancers independently of tumor stage [293]. Reduced levels of T-cadherin mRNA correlated with larger tumor size, lymph node metastasis, invasion, poor differentiation and higher TNM stage [298], while high T-cadherin expression was associated with increased overall survival of gastric cancer patients [298-300]. Overexpression of *CDH13* gene in gastric cell lines resulted in cell cycle arrest, reduced colony formation, migration, invasion and metastasis. These effects were associated with increased E-cadherin, decreased vimentin levels and reduced Akt and mTOR phosphorylation suggesting that T-cadherin may prevent gastric cancer progression by preventing EMT transition and inhibiting the AKT/mTOR signaling pathway [300, 301].

Aberrant methylation of *CDH13* gene promoter has been detected in duodenal carcinomas [302] and colorectal cancer (CRC) cell lines and specimens including both advanced tumors and early adenomatous lesions [303-307, 308, 309, 310, 311] suggesting that inactivation of *CDH13* gene may play a role in malignant transformation of intestinal cells and multistage intestinal cancer progression. *CDH13* gene methylation correlated with poor differentiation [312] and predicted adverse overall survival in CRC [310, 313] but was more frequent in cancers which did not display metastasis to lymph nodes [314, 315].

APN has been suggested to exert anti-tumorigenic activities in different types of cancer [316]. While there is no direct evidence for pathophysiological importance of T-cadherin interactions with APN in cancer cells, APN has been shown to regulate T-cadherin expression in colon cancer cells [317], and CRC risk was significantly associated with genotypes displaying polymorphisms in both *CDH13* and *ADIPOQ* genes [318].

10.6. Liver Cancer

Expression levels and patterns of T-cadherin on ECs within hepatocellular carcinoma (HCC) tumors have been clearly linked to tumor stage and differentiation status [140], further emphasizing the role for T-cadherin in intratumoral angiogenesis. However, the data on expression and functional effects of T-cadherin in HCC cells are variable. On the one hand and similarly to many other cancers, there is evidence for downregulation of *CDH13* gene expression in cancer tissue and HCC metastases due to genetic and epigenetic modifications [203, 319-322]. In contrast, another study reported that while little or no expression of *CDH13* gene was detected in four of five HCC cell lines, it was strongly upregulated in highly invasive Mahlavu cell line, as well as in approximately 50% of primary human HCC tumor specimens (*vs.* non-malignant liver samples) where it correlated with the loss of E-cadherin expression [323, 324]. Functionally, T-cadherin expression has been reported either to induce G(2)/M cell cycle arrest, inhibit cell proliferation and anchorage-independent growth and increase sensitivity to TNFα-induced apoptosis in HCC cells in a c-Jun-dependent manner [320], or to significantly promote motility and invasion without influencing proliferation [139] suggesting that T-cadherin upregulation with concomitant loss of E-cadherin may reflect EMT transition of HCC cells.

10.7. Pancreatic Cancer

Epigenetic modification of T-cadherin promoter was observed in pancreatic cancer cell lines and in 58% cases of primary pancreatic cancers. *CDH13* gene methylation was detected as early as in stage II cancer and in small tumors (less than 2 cm in diameter) suggesting that T-cadherin loss is an early event in pancreatic cancer progression [325].

10.8. Bladder Cancer

CDH13 gene methylation and/or downregulation of expression in different subtypes of bladder tumors, as well as in patients' serum were significantly associated with risk, high grade, invasion and unfavorable prognosis suggesting that T-cadherin may be a relevant diagnostic and predictive biomarker in bladder cancer [326-330].

10.9. Gallblader Cancer

T-cadherin was found to be specifically expressed in non-invasive foci of gallbladder cancer tissue, contrasting with the expression pattern of transcription factor Zeb1 which has been shown promote cancer cell invasion *via* repression of *CDH13* gene promoter activity [331]. Inhibitory effects of T-cadherin on gallblader tumor progression are mediated by downregulation of Akt3 expression and Akt phosphorylation, as well as through regulation of SET7/9-dependent stabilization of chromatin-bound p53 [332].

10.10. Gynaecologic Cancers

CDH13 gene deletion or hypermethylation of its promoter was reported in ovarian cancer [333-337], in hyperplasia and carcinoma of the endometrium [338, 339], and in cervical cancer [340]. Two studies evaluated potential biomarker value of CDH13 and CDH1 gene methylation in serum samples of cervical cancer patients. In the first, an aberrant methylation of the 5'-region of CDH1 or CDH13 genes in 43% of the patients which correlated with worse disease-free survival was demonstrated [341]. However, in another study a statistically significant higher frequency of DNA-methylation was demonstrated only for CDH1 gene, suggesting that CDH13 DNA-hypermethylation is not a good tool for cervical cancer screening due to low diagnostic specificity and sensitivity [342]. In cervical cancer garcinol was observed to exert tumor suppressive effects *via* T-cadherin-dependent activation of PI3K/Akt signaling pathway [343]. On the other hand, in uterine leiomyoma expression of T-cadherin was significantly higher than in adjacent normal myometrium and correlated with increased levels of basic fibroblast growth factor in tumor tissue [344].

10.11. Head and Neck Cancer

CDH13 gene methylation was detected in the sinonasal cancer [345], nasopharyngeal carcinoma [346] and head and neck SCC [347, 348]. Decreased T-cadherin mRNA expression correlated with advanced clinical stage, higher pathological grade, poor differentiation and worse progression-free survival in oral SCC [349]. T-cadherin overexpression inhibited proliferation of oral SCC cell lines through suppression of the PI3K/AKT/mTOR signaling pathway [349].

10.12. Tumors of the Nervous System

In neuroblastoma T-cadherin has been found to inhibit tumor growth by attenuating proliferative cell response to EGF [350]. *CDH13* gene is methylated in glioblastomas [351], and T-cadherin expression decreases during EMT transition and is linked to autophagy and invasion rates of glioblastoma cells [305, 352]. *CDH13* gene has been found to be a direct repression target of Long Non-Coding RNA H19-derived miR-675 which induces expression invasion of glioma cells [353]. Suzuki and co-authors reported an unexpected effect of T-cadherin that may be relevant for fluorescence-guided resection of gliomas [354]. 5-aminolevulinic acid (5-ALA) is metabolized to protoporphyrin IX (PpIX) that accumulates selectively in the tumor and exhibits strong fluorescence upon excitation, but glioma cells do not always respond to 5-ALA, which can result in incomplete or excessive resection. Fluorescence-negative glioma tumors had higher levels of *CDH13* gene expression, and T-cadherin negatively regulated the 5-ALA metabolic pathway since knockdown of *CDH13* gene in U251 glioma cells resulted in higher PpIX accumulation and enhanced fluorescence [354]. T-cadherin has been suggested to be involved in malignant transformation of astrocytes. Astrocytes from mice heterozygous or homozygous for a targeted mutation in the *Nf1* gene displayed increased cell motility and abnormal actin cytoskeleton organization concomitantly with upregulation of T-cadherin [355]. At the same time T-cadherin induced growth arrest in astrocytes in a p21(CIP1/WAF1)-dependent manner [356]. Induction of p21 in response to T-cadherin-dependent contact inhibition has also been demonstrated in CHO cells [357].

10.13. Blood Cancers

CDH13 gene was frequently methylated in lymphoma [358, 359], chronic lymphocytic leukaemia [360], and in bone marrow from patients with acute myeloid leukaemia and chronic myeloid leukaemia (CML) [361]. Expression of T-cadherin mRNA was decreased in primary and blast crisis CML patients as compared with healthy controls and showed a negative correlation with the presence of BCR/ABL fusion gene [362].

10.14. Osteosarcoma

T-cadherin is present in osteosarcoma cell lines and is strongly expressed in primary and metastatic osteosarcoma lesions suggesting that it may promote osteosarcoma progression and metastasis [363-365]. Estradiol, progesterone and EGF are involved in transcriptional and post-transcriptional regulation of T-cadherin in cultured human osteosarcoma cells [365].

10.15. Pituitary Gland Tumors

Hypermethylation of *CDH13* gene was detected in pituitary adenomas where it correlated with invasive tumor phenotype and aggressive higher grade tumors [366, 367].

10.16. Retinoblastoma

LOH on chromosome 16q and specifically *Cdh13* gene has been reported in retinoblastoma [368]. Other studies, however, found no somatic mutations or differential *CDH13* gene expression between retinoblastoma and normal retina, thus suggesting that inactivation of *CDH13* is not likely to be the target of 16q loss in retinoblastoma [369, 370].

CONCLUSION

Abundant data supports the involvement of T-cadherin in a wide variety of (patho)physiological processes, from embryogenesis to cancer progression. The best elucidated functions of T-cadherin include axon navigation, regulation of angiogenesis and control of vascular and cancer cell proliferation, invasion and differentiation. In spite of recent significant progress and increased scientific interest in T-cadherin, the bulk of available data is descriptive, and experimental investigations which directly demonstrate cellular effects of T-cadherin remain scarce. This is especially true for signaling mechanisms mediating T-cadherin effects. The absence of transmembrane and cytoplasmic domains in T-cadherin molecule implies the existence of membrane adaptors which transmit the signals from the extracellular part of the molecule to its intracellular signaling targets. However, analysis of membrane interactions of GPI-anchored proteins is technically challenging. The presence of the lipid moiety renders these receptors highly dynamic and "promiscuous" allowing for fast movements between different membrane compartments and transient lateral interactions with various signaling partners. This property enables efficient functioning in processes requiring prompt cell responses to a changing microenvironment, such as navigation of polarized protruding structures, remodeling of cellular networks, or sensing the environment during directed cell invasion, *inter alia*. Functional outcome of these dynamic signaling events would highly depend on temporal and cellular states at the time of stimulation, as well as the complement of membrane partners (including particular growth factor receptors) present in any given cell type. This is well-illustrated by the many cases of differential effects of T-cadherin in various tissues and contexts described in this chapter: for example, cell responses to T-cadherin-induced Akt axis signaling include stimulation of proliferation and migration of vascular ECs and SMCs but growth arrest of certain tumor cell types. Novel advanced analytical methods of cell biology and biochemistry are required to better understand the molecular machinery of T-cadherin-dependent signal transduction and to clarify the role for this atypical protein in pathogenesis of human diseases.

ACKNOWLEDGMENTS

The authors are supported by the Swiss Herzkreislauf Stiftung and the Swiss Heart Foundation.

REFERENCES

[1] Gul, I. S., Hulpiau, P., Saeys, Y., and van Roy, F. (2017). Evolution and diversity of cadherins and catenins. *Exp Cell Res,* 358: 3-9.

[2] Hulpiau, P., and van Roy, F. (2009). Molecular evolution of the cadherin superfamily. *The international journal of biochemistry & cell biology,* 41: 349-369.

[3] Horne-Badovinac, S. (2017). Fat-like cadherins in cell migration-leading from both the front and the back. *Curr Opin Cell Biol,* 48: 26-32.

[4] Shi, D., Arata, M., Usui, T., Fujimori, T., and Uemura, T. (2016) Seven-Pass Transmembrane Cadherin CELSRs, and Fat4 and Dchs1 Cadherins: From Planar Cell Polarity to Three-Dimensional Organ Architecture. in *The Cadherin Superfamily: Key Regulators of*

Animal Development and Physiology (Suzuki, S. T., and Hirano, S. eds.), Springer Japan, Tokyo. pp 251-276.

[5] Men Mah, K., and Weiner, J. A. (2016) Clustered Protocadherins. in *The Cadherin Superfamily: Key Regulators of Animal Development and Physiology* (Suzuki, S. T., and Hirano, S. eds.), Springer Japan, Toyko. pp 195-222.

[6] Jontes, D. D. (2016) The Nonclustered Protocadherins. in *The Cadherin Superfamily: Key Regulators of Animal Development and Physiology* (Suzuki, S. T., and Hirano, S. eds.), Springer Japan, Tokyo. pp 222-250.

[7] Imai-Okano, K., and Hirano, S. (2016) Various Atypical Cadherins: T-Cadherin, RET, Calsyntenin, and 7D-Cadherin. in *The Cadherin Superfamily: Key Regulators of Animal Development and Physiology* (Suzuki, S. T., and Hirano, S. eds.), Springer Japan, Tokyo. pp 277-314.

[8] Hulpiau, P., Gul, I. S., and van Roy, F. (2016) Evolution of Cadherins and Associated Catenins. in *The Cadherin Superfamily: Key Regulators of Animal Development and Physiology* (Suzuki, S. T., and Hirano, S. eds.), Springer Japan, Tokyo. pp 13-40.

[9] Gumbiner, B. M. (2016) Classical Cadherins. in *The Cadherin Superfamily: Key Regulators of Animal Development and Physiology* (Suzuki, S. T., and Hirano, S. eds.), Springer Japan, Tokyo. pp 41-70.

[10] El-Amraoui, A., and Petit, C. (2016) Cadherins in the Auditory Sensory Organ. in *The Cadherin Superfamily: Key Regulators of Animal Development and Physiology* (Suzuki, S. T., and Hirano, S. eds.), Springer Japan, Tokyo. pp 341-362.

[11] Chidgey, M., and Garrod, D. (2016) Desomsomal Cadherins. in *The Cadherin Superfamily: Key Regulators of Animal Development and Physiology* (Suzuki, S. T., and Hirano, S. eds.), Springer Japan, Tokyo. pp 159-194.

[12] Brayshaw, L. L., and Price, S. R. (2016) Cadherins in Neural Development. in *The Cadherin Superfamily: Key Regulators of*

Animal Development and Physiology (Suzuki, S. T., and Hirano, S. eds.), Springer Japan, Tokyo. pp 315-340.

[13] Albrecht, L. V., Green, K. J., and Dubash, A. D. (2016) Cadherins in Cancer. in *The Cadherin Superfamily: Key Regulators of Animal Development and Physiology* (Suzuki, S. T., and Hirano, S. eds.), Springer Japan, Tokyo. pp 363-398.

[14] Geldhof, A., and Berx, G. (2013). Cadherins and epithelial-to-mesenchymal transition. *Progress in Molecular Biology and Translational Science,* 116: 317-336.

[15] Harrison, O. J., Bahna, F., Katsamba, P. S., Jin, X., Brasch, J., Vendome, J., Ahlsen, G., Carroll, K. J., Price, S. R., Honig, B., and Shapiro, L. (2010). Two-step adhesive binding by classical cadherins. *Nat Struct Mol Biol,* 17: 348-357.

[16] El-Amraoui, A., and Petit, C. (2010). Cadherins as targets for genetic diseases. *Cold Spring Harb Perspect Biol,* 2: a003095.

[17] Braga, V. (2016). Spatial integration of E-cadherin adhesion, signalling and the epithelial cytoskeleton. *Curr Opin Cell Biol,* 42: 138-145.

[18] Campbell, H. K., Maiers, J. L., and DeMali, K. A. (2017). Interplay between tight junctions & adherens junctions. *Exp Cell Res,* 358: 39-44.

[19] Dorland, Y. L., and Huveneers, S. (2017). Cell-cell junctional mechanotransduction in endothelial remodeling. *Cellular and molecular life sciences : CMLS,* 74: 279-292.

[20] Han, M. K., and de Rooij, J. (2016). Converging and Unique Mechanisms of Mechanotransduction at Adhesion Sites. *Trends in cell biology,* 26: 612-623.

[21] Mege, R. M., and Ishiyama, N. (2017). Integration of Cadherin Adhesion and Cytoskeleton at Adherens Junctions. *Cold Spring Harb Perspect Biol,* 9: pii: a028738.

[22] Mui, K. L., Chen, C. S., and Assoian, R. K. (2016). The mechanical regulation of integrin-cadherin crosstalk organizes cells, signaling and forces. *J Cell Sci,* 129: 1093-1100.

[23] Priest, A. V., Shafraz, O., and Sivasankar, S. (2017). Biophysical basis of cadherin mediated cell-cell adhesion. *Exp Cell Res,* 358: 10-13.

[24] Taneyhill, L. A., and Schiffmacher, A. T. (2017). Should I stay or should I go? Cadherin function and regulation in the neural crest. *Genesis (New York, N.Y. : 2000),* 55: e23028.

[25] West, J. J., and Harris, T. J. (2016). Cadherin Trafficking for Tissue Morphogenesis: Control and Consequences. *Traffic (Copenhagen, Denmark),* 17: 1233-1243.

[26] Zhang, X., Liu, J., Liang, X., Chen, J., Hong, J., Li, L., He, Q., and Cai, X. (2016). History and progression of Fat cadherins in health and disease. *OncoTargets and therapy,* 9: 7337-7343.

[27] van Roy, F. (2014). Beyond E-cadherin: roles of other cadherin superfamily members in cancer. *Nature reviews. Cancer,* 14: 121-134.

[28] Philippova, M., Joshi, M. B., Kyriakakis, E., Pfaff, D., Erne, P., and Resink, T. J. (2009). A guide and guard: the many faces of T-cadherin. *Cell Signal,* 21: 1035-1044.

[29] Angst, B. D., Marcozzi, C., and Magee, A. I. (2001). The cadherin superfamily: diversity in form and function. *J Cell Sci,* 114: 629-641.

[30] Ranscht, B., and Dours-Zimmermann, M. T. (1991). T-cadherin, a novel cadherin cell adhesion molecule in the nervous system lacks the conserved cytoplasmic region. *Neuron,* 7: 391-402.

[31] Dames, S. A., Bang, E., Haussinger, D., Ahrens, T., Engel, J., and Grzesiek, S. (2008). Insights into the low adhesive capacity of human T-cadherin from the NMR structure of Its N-terminal extracellular domain. *J Biol Chem,* 283: 23485-23495.

[32] Ciatto, C., Bahna, F., Zampieri, N., VanSteenhouse, H. C., Katsamba, P. S., Ahlsen, G., Harrison, O. J., Brasch, J., Jin, X., Posy, S., Vendome, J., Ranscht, B., Jessell, T. M., Honig, B., and Shapiro, L. (2010). T-cadherin structures reveal a novel adhesive binding mechanism. *Nat Struct Mol Biol,* 17: 339-347.

[33] Vestal, D. J., and Ranscht, B. (1992). Glycosyl phosphatidylinositol--anchored T-cadherin mediates calcium-dependent, homophilic cell adhesion. *J Cell Biol,* 119: 451-461.

[34] Philippova, M., Ivanov, D., Tkachuk, V., Erne, P., and Resink, T. J. (2003). Polarisation of T-cadherin to the leading edge of migrating vascular cells in vitro: a function in vascular cell motility? *Histochem Cell Biol,* 120: 353-360.

[35] Hebbard, L. W., Garlatti, M., Young, L. J., Cardiff, R. D., Oshima, R. G., and Ranscht, B. (2008). T-cadherin supports angiogenesis and adiponectin association with the vasculature in a mouse mammary tumor model. *Cancer research,* 68: 1407-1416.

[36] Lebreton, S., Zurzolo, C., and Paladino, S. (2018). Organization of GPI-anchored proteins at the cell surface and its physiopathological relevance. *Crit Rev Biochem Mol Biol,* 53: 403-419.

[37] Doyle, D. D., Goings, G. E., Upshaw-Earley, J., Page, E., Ranscht, B., and Palfrey, H. C. (1998). T-cadherin is a major glycophosphoinositol-anchored protein associated with noncaveolar detergent-insoluble domains of the cardiac sarcolemma. *J Biol Chem,* 273: 6937-6943.

[38] Philippova, M. P., Bochkov, V. N., Stambolsky, D. V., Tkachuk, V. A., and Resink, T. J. (1998). T-cadherin and signal-transducing molecules co-localize in caveolin-rich membrane domains of vascular smooth muscle cells. *FEBS Lett,* 429: 207-210.

[39] Balatskaya, M. N., Sharonov, G. V., Baglay, A. I., Rubtsov, Y. P., and Tkachuk, V. A. (2019). Different spatiotemporal organization of GPI-anchored T-cadherin in response to low-density lipoprotein and adiponectin. *Biochim Biophys Acta Gen Subj,* 1863: 129414.

[40] Philippova, M., Ivanov, D., Joshi, M. B., Kyriakakis, E., Rupp, K., Afonyushkin, T., Bochkov, V., Erne, P., and Resink, T. J. (2008). Identification of proteins associating with glycosylphosphatidylinositol- anchored T-cadherin on the surface of vascular endothelial cells: role for Grp78/BiP in T-cadherin-dependent cell survival. *Mol Cell Biol,* 28: 4004-4017.

[41] Joshi, M. B., Ivanov, D., Philippova, M., Erne, P., and Resink, T. J. (2007). Integrin-linked kinase is an essential mediator for T-cadherin-dependent signaling via Akt and GSK3beta in endothelial cells. *Faseb J,* 21: 3083-3095.

[42] Philippova, M., Joshi, M. B., Pfaff, D., Kyriakakis, E., Maslova, K., Erne, P., and Resink, T. J. (2012). T-cadherin attenuates insulin-dependent signalling, eNOS activation, and angiogenesis in vascular endothelial cells. *Cardiovasc Res,* 93: 498-507.

[43] Stambolsky, D. V., Kuzmenko, Y. S., Philippova, M. P., Bochkov, V. N., Bespalova, Z. D., Azmuko, A. A., Kashirina, N. M., Vlasik, T. N., Tkachuk, V. A., and Resink, T. J. (1999). Identification of 130 kDa cell surface LDL-binding protein from smooth muscle cells as a partially processed T-cadherin precursor. *Biochimica et biophysica acta,* 1416: 155-160.

[44] Tkachuk, V. A., Bochkov, V. N., Philippova, M. P., Stambolsky, D. V., Kuzmenko, E. S., Sidorova, M. V., Molokoedov, A. S., Spirov, V. G., and Resink, T. J. (1998). Identification of an atypical lipoprotein-binding protein from human aortic smooth muscle as T-cadherin. *FEBS Lett,* 421: 208-212.

[45] Bochkov, V. N., Tkachuk, V. A., Philippova, M. P., Stambolsky, D. V., Buhler, F. R., and Resink, T. J. (1996). Ligand selectivity of 105 kDa and 130 kDa lipoprotein-binding proteins in vascular-smooth-muscle-cell membranes is unique. *Biochem J,* 317 (Pt 1): 297-304.

[46] Kuzmenko, Y. S., Bochkov, V. N., Philippova, M. P., Tkachuk, V. A., and Resink, T. J. (1994). Characterization of an atypical lipoprotein-binding protein in human aortic media membranes by ligand blotting. *Biochem J,* 303 (Pt 1): 281-287.

[47] Hug, C., Wang, J., Ahmad, N. S., Bogan, J. S., Tsao, T. S., and Lodish, H. F. (2004). T-cadherin is a receptor for hexameric and high-molecular-weight forms of Acrp30/adiponectin. *Proc Natl Acad Sci U S A,* 101: 10308-10313.

[48] Fukuda, S., Kita, S., Obata, Y., Fujishima, Y., Nagao, H., Masuda, S., Tanaka, Y., Nishizawa, H., Funahashi, T., Takagi, J., Maeda, N., and Shimomura, I. (2017). The unique prodomain of T-cadherin

plays a key role in adiponectin binding with the essential extracellular cadherin repeats 1 and 2. *J Biol Chem,* 292: 7840-7849.

[49] Niermann, T., Kern, F., Erne, P., and Resink, T. (2000). The glycosyl phosphatidylinositol anchor of human T-cadherin binds lipoproteins. *Biochem Biophys Res Commun,* 276: 1240-1247.

[50] Parker-Duffen, J. L., Nakamura, K., Silver, M., Kikuchi, R., Tigges, U., Yoshida, S., Denzel, M. S., Ranscht, B., and Walsh, K. (2013). T-cadherin is essential for adiponectin-mediated revascularization. *J Biol Chem,* 288: 24886-24897.

[51] Denzel, M. S., Scimia, M. C., Zumstein, P. M., Walsh, K., Ruiz-Lozano, P., and Ranscht, B. (2010). T-cadherin is critical for adiponectin-mediated cardioprotection in mice. *J Clin Invest,* 120: 4342-4352.

[52] Matsuda, K., Fujishima, Y., Maeda, N., Mori, T., Hirata, A., Sekimoto, R., Tsushima, Y., Masuda, S., Yamaoka, M., Inoue, K., Nishizawa, H., Kita, S., Ranscht, B., Funahashi, T., and Shimomura, I. (2015). Positive feedback regulation between adiponectin and T-cadherin impacts adiponectin levels in tissue and plasma of male mice. *Endocrinology,* 156: 934-946.

[53] Fujishima, Y., Maeda, N., Matsuda, K., Masuda, S., Mori, T., Fukuda, S., Sekimoto, R., Yamaoka, M., Obata, Y., Kita, S., Nishizawa, H., Funahashi, T., Ranscht, B., and Shimomura, I. (2017). Adiponectin association with T-cadherin protects against neointima proliferation and atherosclerosis. *Faseb J,* 31: 1571-1583.

[54] Tanaka, Y., Kita, S., Nishizawa, H., Fukuda, S., Fujishima, Y., Obata, Y., Nagao, H., Masuda, S., Nakamura, Y., Shimizu, Y., Mineo, R., Natsukawa, T., Funahashi, T., Ranscht, B., Fukada, S. I., Maeda, N., and Shimomura, I. (2019). Adiponectin promotes muscle regeneration through binding to T-cadherin. *Sci Rep,* 9: 16.

[55] Van Battum, E. Y., Brignani, S., and Pasterkamp, R. J. (2015). Axon guidance proteins in neurological disorders. *Lancet Neurol,* 14: 532-546.

[56] Ranscht, B., and Bronner-Fraser, M. (1991). T-cadherin expression alternates with migrating neural crest cells in the trunk of the avian embryo. *Development,* 111: 15-22.

[57] Fredette, B. J., and Ranscht, B. (1994). T-cadherin expression delineates specific regions of the developing motor axon-hindlimb projection pathway. *The Journal of neuroscience: the official journal of the Society for Neuroscience,* 14: 7331-7346.

[58] Fredette, B. J., Miller, J., and Ranscht, B. (1996). Inhibition of motor axon growth by T-cadherin substrata. *Development,* 122: 3163-3171.

[59] Hayano, Y., Zhao, H., Kobayashi, H., Takeuchi, K., Norioka, S., and Yamamoto, N. (2014). The role of T-cadherin in axonal pathway formation in neocortical circuits. *Development,* 141: 4784-4793.

[60] Miskevich, F., Zhu, Y., Ranscht, B., and Sanes, J. R. (1998). Expression of multiple cadherins and catenins in the chick optic tectum. *Mol Cell Neurosci,* 12: 240-255.

[61] Matsunaga, E., Nambu, S., Oka, M., and Iriki, A. (2015). Complex and dynamic expression of cadherins in the embryonic marmoset cerebral cortex. *Dev Growth Differ,* 57: 474-483.

[62] Thorgeirsson, T. E., Gudbjartsson, D. F., Surakka, I., Vink, J. M., Amin, N., Geller, F., Sulem, P., Rafnar, T., Esko, T., Walter, S., Gieger, C., Rawal, R., Mangino, M., Prokopenko, I., Magi, R., Keskitalo, K., Gudjonsdottir, I. H., Gretarsdottir, S., Stefansson, H., Thompson, J. R., Aulchenko, Y. S., Nelis, M., Aben, K. K., den Heijer, M., Dirksen, A., Ashraf, H., Soranzo, N., Valdes, A. M., Steves, C., Uitterlinden, A. G., Hofman, A., Tonjes, A., Kovacs, P., Hottenga, J. J., Willemsen, G., Vogelzangs, N., Doring, A., Dahmen, N., Nitz, B., Pergadia, M. L., Saez, B., De Diego, V., Lezcano, V., Garcia-Prats, M. D., Ripatti, S., Perola, M., Kettunen, J., Hartikainen, A. L., Pouta, A., Laitinen, J., Isohanni, M., Huei-Yi, S., Allen, M., Krestyaninova, M., Hall, A. S., Jones, G. T., van Rij, A. M., Mueller, T., Dieplinger, B., Haltmayer, M., Jonsson, S., Matthiasson, S. E., Oskarsson, H., Tyrfingsson, T., Kiemeney, L. A., Mayordomo, J. I., Lindholt, J. S., Pedersen, J. H., Franklin, W. A.,

Wolf, H., Montgomery, G. W., Heath, A. C., Martin, N. G., Madden, P. A. F., Giegling, I., Rujescu, D., Jarvelin, M. R., Salomaa, V., Stumvoll, M., Spector, T. D., Wichmann, H. E., Metspalu, A., Samani, N. J., Penninx, B. W., Oostra, B. A., Boomsma, D. I., Tiemeier, H., van Duijn, C. M., Kaprio, J., Gulcher, J. R., McCarthy, M. I., Peltonen, L., Thorsteinsdottir, U., Stefansson, K., and Consortium, E. (2010). Sequence variants at CHRNB3-CHRNA6 and CYP2A6 affect smoking behavior. *Nat Genet,* 42: 448-U135.

[63] Drgon, T., Montoya, I., Johnson, C., Liu, Q. R., Walther, D., Hamer, D., and Uhl, G. R. (2009). Genome-Wide Association for Nicotine Dependence and Smoking Cessation Success in NIH Research Volunteers. *Mol Med,* 15: 21-27.

[64] Uhl, G. R., Walther, D., Musci, R., Fisher, C., Anthony, J. C., Storr, C. L., Behm, F. M., Eaton, W. W., Ialongo, N., and Rose, J. E. (2014). Smoking quit success genotype score predicts quit success and distinct patterns of developmental involvement with common addictive substances. *Molecular psychiatry,* 19: 50-54.

[65] Treutlein, J., Cichon, S., Ridinger, M., Wodarz, N., Soyka, M., Zill, P., Maier, W., Moessner, R., Gaebel, W., Dahmen, N., Fehr, C., Scherbaum, N., Steffens, M., Ludwig, K. U., Frank, J., Wichmann, H. E., Schreiber, S., Dragano, N., Sommer, W. H., Leonardi-Essmann, F., Lourdusamy, A., Gebicke-Haerter, P., Wienker, T. F., Sullivan, P. F., Nothen, M. M., Kiefer, F., Spanagel, R., Mann, K., and Rietschel, M. (2009). Genome-wide association study of alcohol dependence. *Archives of general psychiatry,* 66: 773-784.

[66] Edwards, A. C., Aliev, F., Bierut, L. J., Bucholz, K. K., Edenberg, H., Hesselbrock, V., Kramer, J., Kuperman, S., Nurnberger, J. I., Jr., Schuckit, M. A., Porjesz, B., and Dick, D. M. (2012). Genome-wide association study of comorbid depressive syndrome and alcohol dependence. *Psychiatric genetics,* 22: 31-41.

[67] Johnson, C., Drgon, T., Liu, Q. R., Walther, D., Edenberg, H., Rice, J., Foroud, T., and Uhl, G. R. (2006). Pooled association genome scanning for alcohol dependence using 104,268 SNPs: Validation

and use to identify alcoholism vulnerability loci in unrelated individuals from the collaborative study on the genetics of alcoholism. *Am J Med Genet B,* 141B: 844-853.

[68] Joslyn, G., Ravindranathan, A., Brush, G., Schuckit, M., and White, R. L. (2010). Human Variation in Alcohol Response Is Influenced by Variation in Neuronal Signaling Genes. *Alcohol Clin Exp Res,* 34: 800-812.

[69] Uhl, G. R., Drgon, T., Liu, Q. R., Johnson, C., Walther, D., Komiyama, T., Harano, M., Sekine, Y., Inada, T., Ozaki, N., Iyo, M., Iwata, N., Yamada, M., Sora, I., Chen, C. K., Liu, H. C., Ujike, H., and Lin, S. K. (2008). Genome-wide association for methamphetamine dependence: convergent results from 2 samples. *Archives of general psychiatry,* 65: 345-355.

[70] Uhl, G. R., Drgon, T., Johnson, C., Li, C. Y., Contoreggi, C., Hess, J., Naiman, D., and Liu, Q. R. (2008). Molecular Genetics of Addiction and Related Heritable Phenotypes Genome-Wide Association Approaches Identify "Connectivity Constellation" and Drug Target Genes with Pleiotropic Effects. *Ann Ny Acad Sci,* 1141: 318-381.

[71] Drgon, T., Johnson, C. A., Nino, M., Drgonova, J., Walther, D. M., and Uhl, G. R. (2011). "Replicated" Genome Wide Association for Dependence on Illegal Substances: Genomic Regions Identified by Overlapping Clusters of Nominally Positive SNPs. *Am J Med Genet B,* 156B: 125-138.

[72] Johnson, C., Drgon, T., Walther, D., and Uhl, G. R. (2011). Genomic Regions Identified by Overlapping Clusters of Nominally-Positive SNPs from Genome-Wide Studies of Alcohol and Illegal Substance Dependence. *PloS one,* 6.

[73] Liu, Q. R., Drgon, T., Johnson, C., Walther, D., Hess, J., and Uhl, G. R. (2006). Addiction molecular genetics: 639,401 SNP whole genome association identifies many "cell adhesion" genes. *Am J Med Genet B,* 141B: 918-925.

[74] Johnson, C., Drgon, T., Liu, Q. R., Zhang, P. W., Walther, D., Li, C. Y., Anthony, J. C., Ding, Y. L., Eaton, W. W., and Uhl, G. R.

(2008). Genome wide association for substance dependence: convergent results from epidemiologic and research volunteer samples. *Bmc Med Genet,* 9.

[75] Hart, A. B., Engelhardt, B. E., Wardle, M. C., Sokoloff, G., Stephens, M., de Wit, H., and Palmer, A. A. (2012). Genome-wide association study of d-amphetamine response in healthy volunteers identifies putative associations, including cadherin 13 (CDH13). *PloS one,* 7: e42646.

[76] Leventhal, A. M., Kirkpatrick, M. G., Pester, M. S., McGeary, J. E., Swift, R. M., Sussman, S., and Kahler, C. W. (2017). Pharmacogenetics of stimulant abuse liability: association of CDH13 variant with amphetamine response in a racially-heterogeneous sample of healthy young adults. *Psychopharmacology,* 234: 307-315.

[77] Lein, E. S., Hawrylycz, M. J., Ao, N., Ayres, M., Bensinger, A., Bernard, A., Boe, A. F., Boguski, M. S., Brockway, K. S., Byrnes, E. J., Chen, L., Chen, L., Chen, T. M., Chin, M. C., Chong, J., Crook, B. E., Czaplinska, A., Dang, C. N., Datta, S., Dee, N. R., Desaki, A. L., Desta, T., Diep, E., Dolbeare, T. A., Donelan, M. J., Dong, H. W., Dougherty, J. G., Duncan, B. J., Ebbert, A. J., Eichele, G., Estin, L. K., Faber, C., Facer, B. A., Fields, R., Fischer, S. R., Fliss, T. P., Frensley, C., Gates, S. N., Glattfelder, K. J., Halverson, K. R., Hart, M. R., Hohmann, J. G., Howell, M. P., Jeung, D. P., Johnson, R. A., Karr, P. T., Kawal, R., Kidney, J. M., Knapik, R. H., Kuan, C. L., Lake, J. H., Laramee, A. R., Larsen, K. D., Lau, C., Lemon, T. A., Liang, A. J., Liu, Y., Luong, L. T., Michaels, J., Morgan, J. J., Morgan, R. J., Mortrud, M. T., Mosqueda, N. F., Ng, L. L., Ng, R., Orta, G. J., Overly, C. C., Pak, T. H., Parry, S. E., Pathak, S. D., Pearson, O. C., Puchalski, R. B., Riley, Z. L., Rockett, H. R., Rowland, S. A., Royall, J. J., Ruiz, M. J., Sarno, N. R., Schaffnit, K., Shapovalova, N. V., Sivisay, T., Slaughterbeck, C. R., Smith, S. C., Smith, K. A., Smith, B. I., Sodt, A. J., Stewart, N. N., Stumpf, K. R., Sunkin, S. M., Sutram, M., Tam, A., Teemer, C. D., Thaller, C., Thompson, C. L., Varnam, L. R., Visel, A., Whitlock, R.

M., Wohnoutka, P. E., Wolkey, C. K., Wong, V. Y., Wood, M., Yaylaoglu, M. B., Young, R. C., Youngstrom, B. L., Yuan, X. F., Zhang, B., Zwingman, T. A., and Jones, A. R. (2007). Genome-wide atlas of gene expression in the adult mouse brain. *Nature,* 445: 168-176.

[78] Drgonova, J., Walther, D., Hartstein, G. L., Bukhari, M. O., Baumann, M. H., Katz, J., Hall, F. S., Arnold, E. R., Flax, S., Riley, A., Rivero, O., Lesch, K. P., Troncoso, J., Ranscht, B., and Uhl, G. R. (2016). Cadherin 13: Human cis-Regulation and Selectively Altered Addiction Phenotypes and Cerebral Cortical Dopamine in Knockout Mice. *Mol Med,* 22: 537-547.

[79] King, C. P., Militello, L., Hart, A., St Pierre, C. L., Leung, E., Versaggi, C. L., Roberson, N., Catlin, J., Palmer, A. A., Richards, J. B., and Meyer, P. J. (2017). Cdh13 and AdipoQ gene knockout alter instrumental and Pavlovian drug conditioning. *Genes Brain Behav,* 16: 686-698.

[80] Albrecht, B., Uebel-von Sandersleben, H., Gevensleben, H., and Rothenberger, A. (2015). Pathophysiology of ADHD and associated problems - starting points for NF interventions? *Front Hum Neurosci,* 9.

[81] Lasky-Su, J., Neale, B. M., Franke, B., Anney, R. J. L., Zhou, K. X., Maller, J. B., Vasquez, A. A., Chen, W., Asherson, P., Buitelaar, J., Banaschewski, T., Ebstein, R., Gill, M., Miranda, A., Mulas, F., Oades, R. D., Roeyers, H., Rothenberger, A., Sergeant, J., Sonuga-Barke, E. J. S., Steinhausen, H. C., Taylor, E., Daly, M., Laird, N., Lange, C., and Faraone, S. V. (2008). Genome-Wide Association Scan of Quantitative Traits for Attention Deficit Hyperactivity Disorder Identifies Novel Associations and Confirms Candidate Gene Associations. *Am J Med Genet B,* 147B: 1345-1354.

[82] Salatino-Oliveira, A., Genro, J. P., Polanczyk, G., Zeni, C., Schmitz, M., Kieling, C., Anselmi, L., Menezes, A. M. B., Barros, F. C., Polina, E. R., Mota, N. R., Grevet, E. H., Bau, C. H. D., Rohde, L. A., and Hutz, M. H. (2015). Cadherin-13 Gene is Associated with

Hyperactive/Impulsive Symptoms in Attention/Deficit Hyperactivity Disorder. *Am J Med Genet B,* 168: 162-169.

[83] Lesch, K. P., Timmesfeld, N., Renner, T. J., Halperin, R., Roser, C., Nguyen, T. T., Craig, D. W., Romanos, J., Heine, M., Meyer, J., Freitag, C., Warnke, A., Romanos, M., Schafer, H., Walitza, S., Reif, A., Stephan, D. A., and Jacob, C. (2008). Molecular genetics of adult ADHD: converging evidence from genome-wide association and extended pedigree linkage studies. *Journal of neural transmission,* 115: 1573-1585.

[84] Neale, B. M., Medland, S., Ripke, S., Anney, R. J., Asherson, P., Buitelaar, J., Franke, B., Gill, M., Kent, L., Holmans, P., Middleton, F., Thapar, A., Lesch, K. P., Faraone, S. V., Daly, M., Nguyen, T. T., Schafer, H., Steinhausen, H. C., Reif, A., Renner, T. J., Romanos, M., Romanos, J., Warnke, A., Walitza, S., Freitag, C., Meyer, J., Palmason, H., Rothenberger, A., Hawi, Z., Sergeant, J., Roeyers, H., Mick, E., Biederman, J., and Group, I. I. C. (2010). Case-control genome-wide association study of attention-deficit/hyperactivity disorder. *Journal of the American Academy of Child and Adolescent Psychiatry,* 49: 906-920.

[85] Zhou, K., Dempfle, A., Arcos-Burgos, M., Bakker, S. C., Banaschewski, T., Biederman, J., Buitelaar, J., Castellanos, F. X., Doyle, A., Ebstein, R. P., Ekholm, J., Forabosco, P., Franke, B., Freitag, C., Friedel, S., Gill, M., Hebebrand, J., Hinney, A., Jacob, C., Lesch, K. P., Loo, S. K., Lopera, F., McCracken, J. T., McGough, J. J., Meyer, J., Mick, E., Miranda, A., Muenke, M., Mulas, F., Nelson, S. F., Nguyen, T. T., Oades, R. D., Ogdie, M. N., Palacio, J. D., Pineda, D., Reif, A., Renner, T. J., Roeyers, H., Romanos, M., Rothenberger, A., Schafer, H., Sergeant, J., Sinke, R. J., Smalley, S. L., Sonuga-Barke, E., Steinhausen, H. C., van der Meulen, E., Walitza, S., Warnke, A., Lewis, C. M., Faraone, S. V., and Asherson, P. (2008). Meta-analysis of genome-wide linkage scans of attention deficit hyperactivity disorder. *American journal of medical genetics. Part B, Neuropsychiatric genetics: the official*

publication of the International Society of Psychiatric Genetics, 147B: 1392-1398.

[86] Lionel, A. C., Crosbie, J., Barbosa, N., Goodale, T., Thiruvahindrapuram, B., Rickaby, J., Gazzellone, M., Carson, A. R., Howe, J. L., Wang, Z., Wei, J., Stewart, A. F., Roberts, R., McPherson, R., Fiebig, A., Franke, A., Schreiber, S., Zwaigenbaum, L., Fernandez, B. A., Roberts, W., Arnold, P. D., Szatmari, P., Marshall, C. R., Schachar, R., and Scherer, S. W. (2011). Rare copy number variation discovery and cross-disorder comparisons identify risk genes for ADHD. *Science translational medicine,* 3: 95ra75.

[87] Franke, B., Neale, B. M., and Faraone, S. V. (2009). Genome-wide association studies in ADHD. *Human genetics,* 126: 13-50.

[88] Arias-Vasquez, A., Altink, M. E., Rommelse, N. N. J., Slaats-Willemse, D. I. E., Buschgens, C. J. M., Fliers, E. A., Faraone, S. V., Sergeant, J. A., Oosterlaan, J., Franke, B., and Buitelaar, J. K. (2011). CDH13 is associated with working memory performance in attention deficit/hyperactivity disorder. *Genes Brain Behav,* 10: 844-851.

[89] Linner, R. K., Biroli, P., Kong, E., Meddens, F. W., Wedow, R., Fontana, M. A., Lebreton, M., Tino, S. P., Abdellaoui, A., Hammerschlag, A. R., Nivard, M. G., Okbay, A., Rietveld, C. A., Timshel, P. N., Trzaskowski, M., Vlaming, R., Zund, C. L., Bao, Y. C., Buzdugan, L., Caplin, A. H., Chen, C. Y., Eibich, P., Fontanillas, P., Gonzalez, J. R., Joshi, P. K., Karhunen, V., Kleinman, A., Levin, R. Z., Lill, C. M., Meddens, G. A., Muntane, G., Sanchez-Roige, S., Rooij, F. J., Taskesen, E., Wu, Y., Zhang, F. T., Agee, M., Alipanahi, B., Bell, R. K., Bryc, K., Elson, S. L., Furlotte, N. A., Huber, K. E., Litterman, N. K., McCreight, J. C., McIntyre, M. H., Mountain, J. L., Northover, C. A. M., Pitts, S. J., Sathirapongsasuti, J. F., Sazonova, O. V., Shelton, J. F., Shringarpure, S., Tian, C., Tung, J. Y., Vacic, V., Wilson, C. H., Agbessi, M., Ahsan, H., Alves, I., Andiappan, A., Awadalla, P., Battle, A., Beutner, F., Bonder, M. J., Boomsma, D. I., Christiansen, M., Claringbould, A., Deelen, P., Esko, T., Fave, M. J., Franke, L., Frayling, T., Gharib, S.

A., Gibson, G., Heijmans, B., Hemani, G., Jansen, R., Kahonen, M., Kalnapenkis, A., Kasela, S., Kettunen, J., Kim, Y., Kirsten, H., Kovacs, P., Krohn, K., Kronberg-Guzman, J., Kukushkina, V., Kutalik, Z., Lee, B., Lehtimaki, T., Loeffler, M., Marigorta, U. M., Metspalu, A., Milani, L., Montgomery, G. W., Mueller-Nurasyid, M., Nauck, M., Penninx, B., Perola, M., Pervjakova, N., Pierce, B., Powell, J., Prokisch, H., Psaty, B. M., Raitakari, O., Ring, S., Ripatti, S., Rotzchke, O., Rueger, S., Saha, A., Scholz, M., Schramm, K., Seppala, I., Stumvoll, M., Sullivan, P., Hoen, P. B., Teumer, A., Thiery, J., Tong, L., Tonjes, A., van Dongen, J., van Meurs, J., Verlouw, J., Visscher, P. M., Volker, U., Vosa, U., Westra, H. J., Yaghootkar, H., Yang, J., Zeng, B., Lee, J. J., Pers, T. H., Turley, P., Chen, G. B., Emilsson, V., Oskarsson, S., Pickrell, J. K., Thom, K., Timshel, P., Ahluwalia, T. S., Bacelis, J., Baumbach, C., Bjornsdottir, G., Brandsma, J. H., Concas, M. P., Derringer, J., Furlotte, N. A., Galesloot, T. E., Girotto, G., Gupta, R., Hall, L. M., Harris, S. E., Hofer, E., Horikoshi, M., Huffman, J. E., Kaasik, K., Kalafati, I. P., Kong, A., Lahti, J., Lee, S. J. D., Leeuw, C., Lind, P. A., Lindgren, K. O., Liu, T., Mangino, M., Marten, J., Mihailov, E., Miller, M. B., Most, P. J. D., Oldmeadow, C., Payton, A., Pervjakova, N., Peyrot, W. J., Qian, Y., Raitakari, O., Rueedi, R., Salvi, E., Schmidt, B., Schraut, K. E., Shi, J., Smith, A. V., Poot, R. A., Pourcain, B., Teumer, A., Thorleifsson, G., Verweij, N., Vuckovic, D., Wellmann, J., Westra, H. J., Yang, J. Y., Zhao, W., Zhu, Z. H., Alizadeh, B. Z., Amin, N., Bakshi, A., Baumeister, S. E., Biino, G., Bonnelykke, K., Boyle, P. A., Campbell, H., Cappuccio, F. P., Davies, G., De Neve, J. E., Deloukas, P., Demuth, I., Ding, J., Eibich, P., Eisele, L., Eklund, N., Evans, D. M., Faul, J. D., Feitosa, M. F., Forstner, A. J., Gandin, I., Gunnarsson, B., Halldorsson, B. V., Harris, T. B., Heath, A. C., Hocking, L. J., Holliday, E. G., Homuth, G., Horan, M. A., Hottenga, J. J., Jager, P. L., Jugessur, A., Kaakinen, M. A., Kahonen, M., Kanoni, S., Keltigangas-Jarvinen, L., Kiemeney, L. A. L. M., Kolcic, I., Koskinen, S., Kraja, A. T., Kroh, M., Kutalik, Z., Latvala, A., Launer, L. J., Lebreton, M. P.,

Levinson, D. F., Lichtenstein, P., Lichtner, P., Liewald, D. C. M., Loukola, A., Madden, P. A., Magi, R., Maki-Opas, T., Marioni, R. E., Marques-Vidal, P., McMahon, G., Meisinger, C., Meitinger, T., Milaneschi, Y., Milani, L., Montgomery, G. W., Myhre, R., Nelson, C. P., Nyholt, D. R., Ollier, W. E. R., Palotie, A., Paternoster, L., Pedersen, N. L., Petrovic, K. E., Porteous, D. J., Raikkonen, K., Ring, S. M., Robino, A., Rostapshova, O., Rudan, I., Rustichini, A., Salomaa, V., Sanders, A. R., Sarin, A. P., Schmidt, H., Scott, R. J., Smith, B. H., Smith, J. A., Staessen, J. A., Steinhagen-Thiessen, E., Strauch, K., Terracciano, A., Tobin, M. D., Ulivi, S., Vaccargiu, S., Quaye, L., Venturini, C., Vinkhuyzen, A. A. E., Volker, U., Voelzke, H., Vonk, J. M., Vozzi, D., Waage, J., Ware, E. B., Willemsen, G., Attia, J. R., Bennett, D. A., Berger, K., Bertram, L., Bisgaard, H., Boomsma, D. I., Borecki, I. B., Bultmann, U., Chabris, C. F., Cucca, F., Cusi, D., Deary, J., Dedoussis, G. V., Duijn, C. M., Eriksson, J. G., Franke, B., Franke, L., Gasparini, P., Gejman, P. V., Gieger, C., Grabe, H. J., Gratten, J., Gudnason, V., Harst, P. D., Hayward, C., Hinds, D. A., Hoffmann, W., Hypponen, E., Iacono, W. G., Jacobsson, B., Jarvelin, M. R., Jockel, K. H., Kaprio, J., Kardia, S. L. R., Lehtimaki, T., Lehrer, S. F., Magnusson, P. K. E., Martin, N. G., Mcgue, M., Metspalu, A., Pendleton, N., Penninx, B., Perola, M., Pirastu, N., Pirastu, M., Polasek, O., Posthuma, D., Power, C., Province, M. A., Samani, N. J., Schlessinger, D., Schmidt, R., Sorensen, T. I. A., Spector, T. D., Stefansson, K., Thorsteinsdottir, U., Thurik, A. R., Timpson, N. J., Tiemeier, H., Tung, J. Y., Uitterlinden, A. G., Vitart, V., Vollenweider, P., Weir, D. R., Wilson, J. F., Wright, A. F., Conley, D. C., Krueger, R. F., Smith, G. D., Laibson, D. I., Medland, S. E., Yang, J., Johannesson, M., Visscher, P. M., Esko, T., Koellinger, P. D., Cesarini, D., Benjamin, D. J., Auton, A., Boardman, J. D., Clark, D. W., Conlin, A., Dolan, C. C., Fischbacher, U., Groenen, P. J. F., Harris, K. M., Hasler, G., Hofman, A., Ikram, M. A., Jain, S., Karlsson, R., Kessler, R. C., Kooyman, M., MacKillop, J., Mannikko, M., Morcillo-Suarez, C., McQueen, M. B., Schmidt, K. M., Smart, M.

C., Sutter, M., Thurik, A. R., Uitterlinden, A. G., White, J., de Wit, H., Yang, J., Bertram, L., Boomsma, D. I., Esko, T., Fehr, E., Hinds, D. A., Johannesson, M., Kumari, M., Laibson, D., Magnusson, P. K. E., Meyer, M. N., Navarro, A., Palmer, A. A., Pers, T. H., Posthuma, D., Schunk, D., Stein, M. B., Svento, R., Tiemeier, H., Timmers, P. R. H. J., Turley, P., Ursano, R. J., Wagner, G. G., Wilson, J. F., Gratten, J., Lee, J. J., Cesarini, D., Benjamin, D. J., Koellinger, P. D., Beauchamp, J. P., Team, a. M. R., Consortium, e., Consortium, I. C., and Con, S. S. G. A. (2019). Genome-wide association analyses of risk tolerance and risky behaviors in over 1 million individuals identify hundreds of loci and shared genetic influences. *Nat Genet,* 51: 245-+.

[90] Tiihonen, J., Rautiainen, M. R., Ollila, H. M., Repo-Tiihonen, E., Virkkunen, M., Palotie, A., Pietilainen, O., Kristiansson, K., Joukamaa, M., Lauerma, H., Saarela, J., Tyni, S., Vartiainen, H., Paananen, J., Goldman, D., and Paunio, T. (2015). Genetic background of extreme violent behavior. *Molecular psychiatry,* 20: 786-792.

[91] Sanders, S. J., Ercan-Sencicek, A. G., Hus, V., Luo, R., Murtha, M. T., Moreno-De-Luca, D., Chu, S. H., Moreau, M. P., Gupta, A. R., Thomson, S. A., Mason, C. E., Bilguvar, K., Celestino-Soper, P. B., Choi, M., Crawford, E. L., Davis, L., Wright, N. R., Dhodapkar, R. M., DiCola, M., DiLullo, N. M., Fernandez, T. V., Fielding-Singh, V., Fishman, D. O., Frahm, S., Garagaloyan, R., Goh, G. S., Kammela, S., Klei, L., Lowe, J. K., Lund, S. C., McGrew, A. D., Meyer, K. A., Moffat, W. J., Murdoch, J. D., O'Roak, B. J., Ober, G. T., Pottenger, R. S., Raubeson, M. J., Song, Y., Wang, Q., Yaspan, B. L., Yu, T. W., Yurkiewicz, I. R., Beaudet, A. L., Cantor, R. M., Curland, M., Grice, D. E., Gunel, M., Lifton, R. P., Mane, S. M., Martin, D. M., Shaw, C. A., Sheldon, M., Tischfield, J. A., Walsh, C. A., Morrow, E. M., Ledbetter, D. H., Fombonne, E., Lord, C., Martin, C. L., Brooks, A. I., Sutcliffe, J. S., Cook, E. H., Jr., Geschwind, D., Roeder, K., Devlin, B., and State, M. W. (2011). Multiple recurrent de novo CNVs, including duplications of the

7q11.23 Williams syndrome region, are strongly associated with autism. *Neuron,* 70: 863-885.

[92] Sanders, S. J., Xin, H., Willsey, A. J., Ercan-Sencicek, A. G., Samocha, K. E., Cicek, A. E., Murtha, M. T., Bal, V. H., Bishop, S. L., Shan, D., Goldberg, A. P., Cai, J. L., Keaney, J. F., Klei, L., Mandell, J. D., Moreno-De-Luca, D., Poultney, C. S., Robinson, E. B., Smith, L., Solli-Nowlan, T., Su, M. Y., Teran, N. A., Walker, M. F., Werling, D. M., Beaudet, A. L., Cantor, R. M., Fombonne, E., Geschwind, D. H., Grice, D. E., Lord, C., Lowe, J. K., Mane, S. M., Martin, D. M., Morrow, E. M., Talkowski, M. E., Sutcliffe, J. S., Walsh, C. A., Yu, T. W., Ledbetter, D. H., Martin, C. L., Cook, E. H., Buxbaum, J. D., Daly, M. J., Devlin, B., Roeder, K., State, M. W., and Consortium, A. S. (2015). Insights into Autism Spectrum Disorder Genomic Architecture and Biology from 71 Risk Loci. *Neuron,* 87: 1215-1233.

[93] Borglum, A. D., Demontis, D., Grove, J., Pallesen, J., Hollegaard, M. V., Pedersen, C. B., Hedemand, A., Mattheisen, M., investigators, G., Uitterlinden, A., Nyegaard, M., Orntoft, T., Wiuf, C., Didriksen, M., Nordentoft, M., Nothen, M. M., Rietschel, M., Ophoff, R. A., Cichon, S., Yolken, R. H., Hougaard, D. M., Mortensen, P. B., and Mors, O. (2014). Genome-wide study of association and interaction with maternal cytomegalovirus infection suggests new schizophrenia loci. *Molecular psychiatry,* 19: 325-333.

[94] Mavroconstanti, T., Johansson, S., Winge, I., Knappskog, P. M., and Haavik, J. (2013). Functional Properties of Rare Missense Variants of Human CDH13 Found in Adult Attention Deficit/Hyperactivity Disorder (ADHD) Patients. *PloS one,* 8.

[95] Forero, A., Rivero, O., Waldchen, S., Ku, H. P., Kiser, D. P., Gartner, Y., Pennington, L. S., Waider, J., Gaspar, P., Jansch, C., Edenhofer, F., Resink, T. J., Blum, R., Sauer, M., and Lesch, K. P. (2017). Cadherin-13 Deficiency Increases Dorsal Raphe 5-HT Neuron Density and Prefrontal Cortex Innervation in the Mouse Brain. *Frontiers in cellular neuroscience,* 11: 307.

[96] Paradis, S., Harrar, D. B., Lin, Y. X., Koon, A. C., Hauser, J. L., Griffith, E. C., Zhu, L., Brass, L. F., Chen, C. F., and Greenberg, M. E. (2007). An RNAi-based approach identifies molecules required for glutamatergic and GABAergic synapse development. *Neuron,* 53: 217-232.

[97] Rivero, O., Selten, M. M., Sich, S., Popp, S., Bacmeister, L., Amendola, E., Negwer, M., Schubert, D., Proft, F., Kiser, D., Schmitt, A. G., Gross, C., Kolk, S. M., Strekalova, T., van den Hove, D., Resink, T. J., Kasri, N. N., and Lesch, K. P. (2015). Cadherin-13, a risk gene for ADHD and comorbid disorders, impacts GABAergic function in hippocampus and cognition. *Transl Psychiat,* 5.

[98] Mavroconstanti, T., Halmoy, A., and Haavik, J. (2014). Decreased serum levels of adiponectin in adult attention deficit hyperactivity disorder. *Psychiat Res,* 216: 123-130.

[99] Kiser, D. P., Popp, S., Schmitt-Bohrer, A. G., Strekalova, T., van den Hove, D. L., Lesch, K. P., and Rivero, O. (2019). Early-life stress impairs developmental programming in Cadherin 13 (CDH13)-deficient mice. *Progress in neuro-psychopharmacology & biological psychiatry,* 89: 158-168.

[100] Philippova, M., Ivanov, D., Joshi, M. B., Kyriakakis, E., Rupp, K., Afonyushkin, T., Bochkov, V., Erne, P., and Resink, T. J. (2008). Identification of proteins associating with glycosylphosphatidylinositol-anchored T-cadherin on the surface of vascular endothelial cells: Role for Grp78/BiP in T-cadherin-dependent cell survival. *Mol Cell Biol,* 28: 4004-4017.

[101] Killen, A. C., Barber, M., Paulin, J. J. W., Ranscht, B., Parnavelas, J. G., and Andrews, W. D. (2017). Protective role of Cadherin 13 in interneuron development. *Brain Struct Funct,* 222: 3567-3585.

[102] Tantra, M., Guo, L., Kim, J., Zainolabidin, N., Eulenburg, V., Augustine, G. J., and Chen, A. I. (2018). Conditional deletion of Cadherin 13 perturbs Golgi cells and disrupts social and cognitive behaviors. *Genes Brain Behav,* 17.

[103] Ivanov, D., Philippova, M., Antropova, J., Gubaeva, F., Iljinskaya, O., Tararak, E., Bochkov, V., Erne, P., Resink, T., and Tkachuk, V. (2001). Expression of cell adhesion molecule T-cadherin in the human vasculature. *Histochem Cell Biol,* 115: 231-242.

[104] Kostopoulos, C. G., Spiroglou, S. G., Varakis, J. N., Apostolakis, E., and Papadaki, H. H. (2014). Adiponectin/T-cadherin and apelin/APJ expression in human arteries and periadventitial fat: implication of local adipokine signaling in atherosclerosis? *Cardiovasc Pathol,* 23: 131-138.

[105] Org, E., Eyheramendy, S., Juhanson, P., Gieger, C., Lichtner, P., Klopp, N., Veldre, G., Doring, A., Viigimaa, M., Sober, S., Tomberg, K., Eckstein, G., KORA, Kelgo, P., Rebane, T., Shaw-Hawkins, S., Howard, P., Onipinla, A., Dobson, R. J., Newhouse, S. J., Brown, M., Dominiczak, A., Connell, J., Samani, N., Farrall, M., BRIGHT, Caulfield, M. J., Munroe, P. B., Illig, T., Wichmann, H. E., Meitinger, T., and Laan, M. (2009). Genome-wide scan identifies CDH13 as a novel susceptibility locus contributing to blood pressure determination in two European populations. *Hum Mol Genet,* 18: 2288-2296.

[106] Vargas-Alarcon, G., Martinez-Rodriguez, N., Velazquez-Cruz, R., Perez-Mendez, O., Posadas-Sanchez, R., Posadas-Romero, C., Pena-Duque, M. A., Martinez-Rios, M. A., Ramirez-Fuentes, S., and Fragoso, J. M. (2017). The T>A (rs11646213) gene polymorphism of cadherin-13 (CDH13) gene is associated with decreased risk of developing hypertension in Mexican population. *Immunobiology,* 222: 973-978.

[107] Wan, J. P., Zhao, H., Li, T., Li, C. Z., Wang, X. T., and Chen, Z. J. (2013). The common variant rs11646213 is associated with preeclampsia in Han Chinese women. *PloS one,* 8: e71202.

[108] Kim, H. J., Seo, Y. S., Sung, J., Son, H. Y., Yun, J. M., Kwon, H., Cho, B., Kim, J. I., and Park, J. H. (2019). Interactions of CDH13 gene polymorphisms and ambient PM10 air pollution exposure with blood pressure and hypertension in Korean men. *Chemosphere,* 218: 292-298.

[109] Dong, C., Beecham, A., Wang, L., Slifer, S., Wright, C. B., Blanton, S. H., Rundek, T., and Sacco, R. L. (2011). Genetic loci for blood lipid levels identified by linkage and association analyses in Caribbean Hispanics. *Journal of lipid research,* 52: 1411-1419.

[110] An, P., Straka, R. J., Pollin, T. I., Feitosa, M. F., Wojczynski, M. K., Daw, E. W., O'Connell, J. R., Gibson, Q., Ryan, K. A., Hopkins, P. N., Tsai, M. Y., Lai, C. Q., Province, M. A., Ordovas, J. M., Shuldiner, A. R., Arnett, D. K., and Borecki, I. B. (2014). Genome-wide association studies identified novel loci for non-high-density lipoprotein cholesterol and its postprandial lipemic response. *Human genetics,* 133: 919-930.

[111] Bressler, J., Folsom, A. R., Couper, D. J., Volcik, K. A., and Boerwinkle, E. (2010). Genetic variants identified in a European genome-wide association study that were found to predict incident coronary heart disease in the atherosclerosis risk in communities study. *American journal of epidemiology,* 171: 14-23.

[112] Lee, J. H., Shin, D. J., Park, S., Kang, S. M., Jang, Y., and Lee, S. H. (2013). Association between CDH13 variants and cardiometabolic and vascular phenotypes in a Korean population. *Yonsei medical journal,* 54: 1305-1312.

[113] Chotchaeva, F. R., Balatskiy, A. V., Larisa M Samokhodskaya, L.-M.-., Vsevolod A Tkachuk, V. A., A, V., and Sadovnichiy, V. A. (2016). Association between T-cadherin gene (CDH13) variants and severity of coronary heart disease manifestation. *International journal of clinical and experimental medicine,* 9: 4059-4064.

[114] Chen, L., Sun, K. X., Juan, J., Fang, K., Liu, K., Wang, X. Y., Wang, L., Yang, C., Liu, X. Q., Li, J., Tang, X., Wu, Y. Q., Qin, X. Y., Wu, T., Chen, D. F., and Hu, Y. H. (2017). CDH13 Genetic Polymorphisms, Adiponectin and Ischemic Stroke: a Chinese Family-based Sib-pair Study. *Biomed Environ Sci,* 30: 35-43.

[115] Chung, C. M., Lin, T. H., Chen, J. W., Leu, H. B., Yang, H. C., Ho, H. Y., Ting, C. T., Sheu, S. H., Tsai, W. C., Chen, J. H., Lin, S. J., Chen, Y. T., and Pan, W. H. (2011). A genome-wide association

study reveals a quantitative trait locus of adiponectin on CDH13 that predicts cardiometabolic outcomes. *Diabetes,* 60: 2417-2423.

[116] Morisaki, H., Yamanaka, I., Iwai, N., Miyamoto, Y., Kokubo, Y., Okamura, T., Okayama, A., and Morisaki, T. (2012). CDH13 gene coding T-cadherin influences variations in plasma adiponectin levels in the Japanese population. *Human mutation,* 33: 402-410.

[117] Balatsky, A. V., Konovalov, D. Y., Samokhodskaya, L. M., Kochegura, T. N., Rubina, K. A., and Tkachuk, V. A. (2015). [Single Nucleotide Polymorphisms in T-Cadherin Gene (CDH13) Have Cumulative Effect on Body Mass in Patients With Ischemic Heart Disease]. *Kardiologiia,* 55: 12-15.

[118] Ling, H., Waterworth, D. M., Stirnadel, H. A., Pollin, T. I., Barter, P. J., Kesaniemi, Y. A., Mahley, R. W., McPherson, R., Waeber, G., Bersot, T. P., Cohen, J. C., Grundy, S. M., Mooser, V. E., and Mitchell, B. D. (2009). Genome-wide linkage and association analyses to identify genes influencing adiponectin levels: the GEMS Study. *Obesity,* 17: 737-744.

[119] Li, Y., Li, C., Yang, Y., Shi, L., Tao, W., Liu, S., Yang, M., Li, X., Yao, Y., and Xiao, C. (2017). The association of six single nucleotide polymorphisms and their haplotypes in CDH13 with T2DM in a Han Chinese population. *Medicine,* 96: e7063.

[120] Nicolas, A., Aubert, R., Bellili-Munoz, N., Balkau, B., Bonnet, F., Tichet, J., Velho, G., Marre, M., Roussel, R., and Fumeron, F. (2017). T-cadherin gene variants are associated with type 2 diabetes and the Fatty Liver Index in the French population. *Diabetes Metab,* 43: 33-39.

[121] Nicolas, A., Mohammedi, K., Bastard, J. P., Fellahi, S., Bellili-Munoz, N., Roussel, R., Hadjadj, S., Marre, M., Velho, G., and Fumeron, F. (2017). T-cadherin gene variants are associated with nephropathy in subjects with type 1 diabetes. *Nephrol Dial Transplant,* 32: 1987-1993.

[122] Gao, H., Kim, Y. M., Chen, P., Igase, M., Kawamoto, R., Kim, M. K., Kohara, K., Lee, J., Miki, T., Ong, R. T., Onuma, H., Osawa, H., Sim, X., Teo, Y. Y., Tabara, Y., Tai, E. S., and van Dam, R. M.

(2013). Genetic variation in CDH13 is associated with lower plasma adiponectin levels but greater adiponectin sensitivity in East Asian populations. *Diabetes,* 62: 4277-4283.

[123] Li, Y., Yang, Y., Yao, Y., Li, X., Shi, L., Zhang, Y., Xiong, Y., Yan, M., Yao, Y., and Xiao, C. (2014). Association study of ARL15 and CDH13 with T2DM in a Han Chinese population. *Int J Med Sci,* 11: 522-527.

[124] Jee, S. H., Sull, J. W., Lee, J. E., Shin, C., Park, J., Kimm, H., Cho, E. Y., Shin, E. S., Yun, J. E., Park, J. W., Kim, S. Y., Lee, S. J., Jee, E. J., Baik, I., Kao, L., Yoon, S. K., Jang, Y., and Beaty, T. H. (2010). Adiponectin concentrations: a genome-wide association study. *Am J Hum Genet,* 87: 545-552.

[125] Teng, M. S., Wu, S. M., Hsu, L. A., Chou, H. H., and Ko, Y. L. (2015). Differential Associations between CDH13 Genotypes, Adiponectin Levels, and Circulating Levels of Cellular Adhesive Molecules. *Mediat Inflamm,*

[126] Teng, M. S., Hsu, L. A., Wu, S. M., Sun, Y. C., Juan, S. H., and Ko, Y. L. (2015). Association of CDH13 Genotypes/Haplotypes with Circulating Adiponectin Levels, Metabolic Syndrome, and Related Metabolic Phenotypes: The Role of the Suppression Effect. *PloS one,* 10.

[127] Putku, M., Kals, M., Inno, R., Kasela, S., Org, E., Kozich, V., Milani, L., and Laan, M. (2015). CDH13 promoter SNPs with pleiotropic effect on cardiometabolic parameters represent methylation QTLs. *Human genetics,* 134: 291-303.

[128] Kitamoto, A., Kitamoto, T., Nakamura, T., Matsuo, T., Nakata, Y., Hyogo, H., Ochi, H., Kamohara, S., Miyatake, N., Kotani, K., Mineo, I., Wada, J., Ogawa, Y., Yoneda, M., Nakajima, A., Funahashi, T., Miyazaki, S., Tokunaga, K., Masuzaki, H., Ueno, T., Chayama, K., Hamaguchi, K., Yamada, K., Hanafusa, T., Oikawa, S., Sakata, T., Tanaka, K., Matsuzawa, Y., and Hotta, K. (2016). CDH13 Polymorphisms are Associated with Adiponectin Levels and Metabolic Syndrome Traits Independently of Visceral Fat Mass. *Journal of Atherosclerosis and Thrombosis,* 23: 309-319.

[129] Uetani, E., Tabara, Y., Kawamoto, R., Onuma, H., Kohara, K., Osawa, H., and Miki, T. (2014). CDH13 Genotype-Dependent Association of High-Molecular Weight Adiponectin With All-Cause Mortality: The J-SHIPP Study. *Diabetes Care,* 37: 396-401.

[130] Fava, C., Danese, E., Montagnana, M., Sjogren, M., Almgren, P., Guidi, G. C., Hedblad, B., Engstrom, G., Lechi, A., Minuz, P., and Melander, O. (2011). A variant upstream of the CDH13 adiponectin receptor gene and metabolic syndrome in Swedes. *Am J Cardiol,* 108: 1432-1437.

[131] Bag, S., and Anbarasu, A. (2015). Revealing the Strong Functional Association of adipor2 and cdh13 with adipoq: A Gene Network Study. *Cell biochemistry and biophysics,* 71: 1445-1456.

[132] Deanfield, J. E., Halcox, J. P., and Rabelink, T. J. (2007). Endothelial function and dysfunction: testing and clinical relevance. *Circulation,* 115: 1285-1295.

[133] Greyling, A., Hopman, M. T., and Thijssen, D. H. J. (2015) Endothelial Function in Health and Disease. in *Arterial Disorders* (Berbar, A., and Mancia, G. eds.), Springer, Cham. pp 161-173.

[134] Rajendran, P., Rengarajan, T., Thangavel, J., Nishigaki, Y., Sakthisekaran, D., Sethi, G., and Nishigaki, I. (2013). The vascular endothelium and human diseases. *Int J Biol Sci,* 9: 1057-1069.

[135] Joshi, M. B., Philippova, M., Ivanov, D., Allenspach, R., Erne, P., and Resink, T. J. (2005). T-cadherin protects endothelial cells from oxidative stress-induced apoptosis. *Faseb J,* 19: 1737-1739.

[136] Kudrjashova, E., Bashtrikov, P., Bochkov, V., Parfyonova, Y., Tkachuk, V., Antropova, J., Iljinskaya, O., Tararak, E., Erne, P., Ivanov, D., Philippova, M., and Resink, T. J. (2002). Expression of adhesion molecule T-cadherin is increased during neointima formation in experimental restenosis. *Histochem Cell Biol,* 118: 281-290.

[137] Wyder, L., Vitaliti, A., Schneider, H., Hebbard, L. W., Moritz, D. R., Wittmer, M., Ajmo, M., and Klemenz, R. (2000). Increased expression of H/T-cadherin in tumor-penetrating blood vessels. *Cancer research,* 60: 4682-4688.

[138] Oshima, R. G., Lesperance, J., Munoz, V., Hebbard, L., Ranscht, B., Sharan, N., Muller, W. J., Hauser, C. A., and Cardiff, R. D. (2004). Angiogenic acceleration of Neu induced mammary tumor progression and metastasis. *Cancer research,* 64: 169-179.

[139] Riou, P., Saffroy, R., Chenailler, C., Franc, B., Gentile, C., Rubinstein, E., Resink, T., Debuire, B., Piatier-Tonneau, D., and Lemoine, A. (2006). Expression of T-cadherin in tumor cells influences invasive potential of human hepatocellular carcinoma. *Faseb J,* 20: 2291-2301.

[140] Adachi, Y., Takeuchi, T., Sonobe, H., and Ohtsuki, Y. (2006). An adiponectin receptor, T-cadherin, was selectively expressed in intratumoral capillary endothelial cells in hepatocellular carcinoma: possible cross talk between T-cadherin and FGF-2 pathways. *Virchows Arch,* 448: 311-318.

[141] Chan, D. W., Lee, J. M., Chan, P. C., and Ng, I. O. (2008). Genetic and epigenetic inactivation of T-cadherin in human hepatocellular carcinoma cells. *Int J Cancer,* 123: 1043-1052.

[142] Ivanov, D., Philippova, M., Tkachuk, V., Erne, P., and Resink, T. (2004). Cell adhesion molecule T-cadherin regulates vascular cell adhesion, phenotype and motility. *Exp Cell Res,* 293: 207-218.

[143] Joshi, M. B., Ivanov, D., Philippova, M., Kyriakakis, E., Erne, P., and Resink, T. J. (2008). A requirement for thioredoxin in redox-sensitive modulation of T-cadherin expression in endothelial cells. *Biochem J,* 416: 271-280.

[144] Philippova, M., Ivanov, D., Allenspach, R., Takuwa, Y., Erne, P., and Resink, T. (2005). RhoA and Rac mediate endothelial cell polarization and detachment induced by T-cadherin. *Faseb J,* 19: 588-590.

[145] Philippova, M., Banfi, A., Ivanov, D., Gianni-Barrera, R., Allenspach, R., Erne, P., and Resink, T. (2006). Atypical GPI-anchored T-cadherin stimulates angiogenesis *in vitro* and *in vivo*. *Arterioscler Thromb Vasc Biol,* 26: 2222-2230.

[146] Philippova, M., Suter, Y., Toggweiler, S., Schoenenberger, A. W., Joshi, M. B., Kyriakakis, E., Erne, P., and Resink, T. J. (2011). T-

cadherin is present on endothelial microparticles and is elevated in plasma in early atherosclerosis. *Eur Heart J,* 32: 760-771.
[147] Kyriakakis, E., Philippova, M., Joshi, M. B., Pfaff, D., Bochkov, V., Afonyushkin, T., Erne, P., and Resink, T. J. (2010). T-cadherin attenuates the PERK branch of the unfolded protein response and protects vascular endothelial cells from endoplasmic reticulum stress-induced apoptosis. *Cell Signal,* 22: 1308-1316.
[148] Joshi, M. B., Kyriakakis, E., Pfaff, D., Rupp, K., Philippova, M., Erne, P., and Resink, T. J. (2009). Extracellular cadherin repeat domains EC1 and EC5 of T-cadherin are essential for its ability to stimulate angiogenic behavior of endothelial cells. *Faseb J,* 23: 4011-4021.
[149] Nomura-Nakayama, K., Adachi, H., Miyatake, N., Hayashi, N., Fujimoto, K., Yamaya, H., and Yokoyama, H. (2018). High molecular weight adiponectin inhibits vascular calcification in renal allograft recipients. *PloS one,* 13.
[150] Wang, Y., Lam, K. S., Xu, J. Y., Lu, G., Xu, L. Y., Cooper, G. J., and Xu, A. (2005). Adiponectin inhibits cell proliferation by interacting with several growth factors in an oligomerization-dependent manner. *J Biol Chem,* 280: 18341-18347.
[151] Ouchi, N., Kobayashi, H., Kihara, S., Kumada, M., Sato, K., Inoue, T., Funahashi, T., and Walsh, K. (2004). Adiponectin stimulates angiogenesis by promoting cross-talk between AMP-activated protein kinase and Akt signaling in endothelial cells. *J Biol Chem,* 279: 1304-1309.
[152] Herring, J. M., McMichael, M. A., and Smith, S. A. (2013). Microparticles in health and disease. *J Vet Intern Med,* 27: 1020-1033.
[153] Hugel, B., Martinez, M. C., Kunzelmann, C., and Freyssinet, J. M. (2005). Membrane microparticles: two sides of the coin. *Physiology (Bethesda),* 20: 22-27.
[154] Barteneva, N. S., Fasler-Kan, E., Bernimoulin, M., Stern, J. N., Ponomarev, E. D., Duckett, L., and Vorobjev, I. A. (2013). Circulating microparticles: square the circle. *BMC Cell Biol,* 14: 23.

[155] Chironi, G. N., Boulanger, C. M., Simon, A., Dignat-George, F., Freyssinet, J. M., and Tedgui, A. (2009). Endothelial microparticles in diseases. *Cell Tissue Res,* 335: 143-151.

[156] Constans, J., and Conri, C. (2006). Circulating markers of endothelial function in cardiovascular disease. *Clin Chim Acta,* 368: 33-47.

[157] Gomez, D., and Owens, G. K. (2012). Smooth muscle cell phenotypic switching in atherosclerosis. *Cardiovasc Res,* 95: 156-164.

[158] Miano, J. M. (2010). Vascular smooth muscle cell differentiation-2010. *J Biomed Res,* 24: 169-180.

[159] Rensen, S. S., Doevendans, P. A., and van Eys, G. J. (2007). Regulation and characteristics of vascular smooth muscle cell phenotypic diversity. *Neth Heart J,* 15: 100-108.

[160] Rzucidlo, E. M., Martin, K. A., and Powell, R. J. (2007). Regulation of vascular smooth muscle cell differentiation. *J Vasc Surg,* 45 Suppl A: A25-32.

[161] Shi, N., and Chen, S. Y. (2015). Smooth Muscle Cell Differentiation: Model Systems, Regulatory Mechanisms, and Vascular Diseases. *Journal of cellular physiology,*

[162] Qin, M., Zeng, Z., Zheng, J., Shah, P. K., Schwartz, S. M., Adams, L. D., and Sharifi, B. G. (2003). Suppression subtractive hybridization identifies distinctive expression markers for coronary and internal mammary arteries. *Arterioscler Thromb Vasc Biol,* 23: 425-433.

[163] Takeuchi, T., Adachi, Y., Ohtsuki, Y., and Furihata, M. (2007). Adiponectin receptors, with special focus on the role of the third receptor, T-cadherin, in vascular disease. *Med Mol Morphol,* 40: 115-120.

[164] Collot-Teixeira, S., McGregor, J. L., Morser, K., Chalabreysse, L., McDermott-Roe, C., Cerutti, C., Guzman, A., Michel, J. B., Boissonnat, P., Sebbag, L., Thivolet-Bejui, F., Bricca, G., Obadia, J. F., and Roussoulieres, A. (2010). T-cadherin expression in cardiac allograft vasculopathy: bench to bedside translational investigation.

The Journal of heart and lung transplantation: the official publication of the International Society for Heart Transplantation, 29: 792-799.

[165] Frismantiene, A., Pfaff, D., Frachet, A., Coen, M., Joshi, M. B., Maslova, K., Bochaton-Piallat, M. L., Erne, P., Resink, T. J., and Philippova, M. (2014). Regulation of contractile signaling and matrix remodeling by T-cadherin in vascular smooth muscle cells: constitutive and insulin-dependent effects. *Cell Signal,* 26: 1897-1908.

[166] Frismantiene, A., Dasen, B., Pfaff, D., Erne, P., Resink, T. J., and Philippova, M. (2016). T-cadherin promotes vascular smooth muscle cell dedifferentiation via a GSK3beta-inactivation dependent mechanism. *Cell Signal,* 28: 516-530.

[167] Miano, J. M., Long, X. C., and Fujiwara, K. (2007). Serum response factor: master regulator of the actin cytoskeleton and contractile apparatus. *Am J Physiol-Cell Ph,* 292: C70-C81.

[168] Parmacek, M. S. (2007). Myocardin-related transcription factors: critical coactivators regulating cardiovascular development and adaptation. *Circ Res,* 100: 633-644.

[169] Parmacek, M. S. (2008). Myocardin: dominant driver of the smooth muscle cell contractile phenotype. *Arterioscler Thromb Vasc Biol,* 28: 1416-1417.

[170] Ivanov, D., Philippova, M., Allenspach, R., Erne, P., and Resink, T. (2004). T-cadherin upregulation correlates with cell-cycle progression and promotes proliferation of vascular cells. *Cardiovasc Res,* 64: 132-143.

[171] Frismantiene, A., Kyriakakis, E., Dasen, B., Erne, P., Resink, T. J., and Philippova, M. (2017). Actin cytoskeleton regulates functional anchorage-migration switch during T-cadherin-induced phenotype modulation of vascular smooth muscle cells. *Cell Adh Migr*: 1-17.

[172] Kyriakakis, E., Frismantiene, A., Dasen, B., Pfaff, D., Rivero, O., Lesch, K. P., Erne, P., Resink, T. J., and Philippova, M. (2017). T-cadherin promotes autophagy and survival in vascular smooth

muscle cells through MEK1/2/Erk1/2 axis activation. *Cell Signal,* 35: 163-175.

[173] Salabei, J. K., Cummins, T. D., Singh, M., Jones, S. P., Bhatnagar, A., and Hill, B. G. (2013). PDGF-mediated autophagy regulates vascular smooth muscle cell phenotype and resistance to oxidative stress. *Biochem J,* 451: 375-388.

[174] Salabei, J. K., and Hill, B. G. (2013). Implications of autophagy for vascular smooth muscle cell function and plasticity. *Free Radic Biol Med,* 65: 693-703.

[175] Frismantiene, A., Philippova, M., Erne, P., and Resink, T. J. (2018). Smooth muscle cell-driven vascular diseases and molecular mechanisms of VSMC plasticity. *Cell Signal,* 52: 48-64.

[176] Frismantiene, A., Philippova, M., Erne, P., and Resink, T. J. (2018). Cadherins in vascular smooth muscle cell (patho)biology: Quid nos scimus? *Cell Signal,* 45: 23-42.

[177] Frismantiene, A., Kyriakakis, E., Dasen, B., Erne, P., Resink, T. J., and Philippova, M. (2018). Actin cytoskeleton regulates functional anchorage-migration switch during T-cadherin-induced phenotype modulation of vascular smooth muscle cells. *Cell Adh Migr,* 12: 69-85.

[178] Ference, B. A., Ginsberg, H. N., Graham, I., Ray, K. K., Packard, C. J., Bruckert, E., Hegele, R. A., Krauss, R. M., Raal, F. J., Schunkert, H., Watts, G. F., Boren, J., Fazio, S., Horton, J. D., Masana, L., Nicholls, S. J., Nordestgaard, B. G., van de Sluis, B., Taskinen, M. R., Tokgozoglu, L., Landmesser, U., Laufs, U., Wiklund, O., Stock, J. K., Chapman, M. J., and Catapano, A. L. (2017). Low-density lipoproteins cause atherosclerotic cardiovascular disease. 1. Evidence from genetic, epidemiologic, and clinical studies. A consensus statement from the European Atherosclerosis Society Consensus Panel. *Eur Heart J,* 38: 2459-2472.

[179] Geovanini, G. R., and Libby, P. (2018). Atherosclerosis and inflammation: overview and updates. *Clin Sci (Lond),* 132: 1243-1252.

[180] Rubina, K., Talovskaya, E., Cherenkov, V., Ivanov, D., Stambolsky, D., Storozhevykh, T., Pinelis, V., Shevelev, A., Parfyonova, Y., Resink, T., Erne, P., and Tkachuk, V. (2005). LDL induces intracellular signalling and cell migration via atypical LDL-binding protein T-cadherin. *Mol Cell Biochem,* 273: 33-41.

[181] Kipmen-Korgun, D., Osibow, K., Zoratti, C., Greilberger, J., Kostner, G. M., Juergens, G., and Graier, W. F. (2006). T-cadherin mediates low-density lipoprotein-initiated mitogenic signalling. *Febs J,* 273: 94-94.

[182] Kipmen-Korgun, D., Osibow, K., Zoratti, C., Schraml, E., Greilberger, J., Kostner, G. M., Jurgens, G., and Graier, W. F. (2005). T-cadherin mediates low-density lipoprotein-initiated cell proliferation via the Ca2+-tyrosine kinase-Erk1/2 pathway. *J Cardiovasc Pharm,* 45: 418-430.

[183] Bochkov, V., Tkachuk, V., Buhler, F., and Resink, T. (1992). Phosphoinositide and calcium signalling responses in smooth muscle cells: comparison between lipoproteins, Ang II, and PDGF. *Biochem Biophys Res Commun,* 188: 1295-1304.

[184] Sachinidis, A., Seewald, S., Epping, P., Seul, C., Ko, Y., and Vetter, H. (1997). The growth-promoting effect of low-density lipoprotein may Be mediated by a pertussis toxin-sensitive mitogen-activated protein kinase pathway. *Mol Pharmacol,* 52: 389-397.

[185] Sachinidis, A., Mengden, T., Locher, R., Brunner, C., and Vetter, W. (1990). Novel cellular activities for low density lipoprotein in vascular smooth muscle cells. *Hypertension,* 15: 704-711.

[186] Scott-Burden, T., Resink, T. J., Hahn, A. W., Baur, U., Box, R. J., and Buhler, F. R. (1989). Induction of growth-related metabolism in human vascular smooth muscle cells by low density lipoprotein. *J Biol Chem,* 264: 12582-12589.

[187] Locher, R., Brandes, R. P., Vetter, W., and Barton, M. (2002). Native LDL induces proliferation of human vascular smooth muscle cells via redox-mediated activation of ERK 1/2 mitogen-activated protein kinases. *Hypertension,* 39: 645-650.

[188] Gouni-Berthold, I., and Sachinidis, A. (2002). Does the coronary risk factor low density lipoprotein alter growth and signaling in vascular smooth muscle cells? *Faseb J,* 16: 1477-1487.

[189] Escudero, C. A., Herlitz, K., Troncoso, F., Guevara, K., Acurio, J., Aguayo, C., Godoy, A. S., and Gonzalez, M. (2017). Pro-angiogenic Role of Insulin: From Physiology to Pathology. *Front Physiol,* 8: 204.

[190] Wang, C. C., Gurevich, I., and Draznin, B. (2003). Insulin affects vascular smooth muscle cell phenotype and migration via distinct signaling pathways. *Diabetes,* 52: 2562-2569.

[191] Xi, G., Shen, X., Wai, C., White, M. F., and Clemmons, D. R. (2019). Hyperglycemia induces vascular smooth muscle cell dedifferentiation by suppressing insulin receptor substrate-1-mediated p53/KLF4 complex stabilization. *J Biol Chem,* 294: 2407-2421.

[192] Tyrberg, B., Miles, P., Azizian, K. T., Denzel, M. S., Nieves, M. L., Monosov, E. Z., Levine, F., and Ranscht, B. (2011). T-cadherin (Cdh13) in association with pancreatic beta-cell granules contributes to second phase insulin secretion. *Islets,* 3: 327-337.

[193] Wang, H., Tao, L., Ambrosio, A., Yan, W., Summer, R., Lau, W. B., Wang, Y., and Ma, X. (2017). T-cadherin deficiency increases vascular vulnerability in T2DM through impaired NO bioactivity. *Cardiovasc Diabetol,* 16: 12.

[194] Wang, L. L., Miller, D., Wanders, D., Nanayakkara, G., Amin, R., Judd, R., Morri-Son, E. E., and Zhong, J. M. (2016). Adiponectin downregulation is associated with volume overload-induced myocyte dysfunction in rats. *Acta Pharmacologica Sinica,* 37: 187-195.

[195] Sideleva, O., Suratt, B. T., Black, K. E., Tharp, W. G., Pratley, R. E., Forgione, P., Dienz, O., Irvin, C. G., and Dixon, A. E. (2012). Obesity and asthma: an inflammatory disease of adipose tissue not the airway. *Am J Respir Crit Care Med,* 186: 598-605.

[196] Yuan, Y. M., Zhang, J. L., Xu, S. C., Ye, R. S., Xu, D., Zhang, Y., Zhang, Y. J., Chen, Y. L., Liu, Y. L., and Su, Z. G. (2016). Genetic

variants of CDH13 determine the susceptibility to chronic obstructive pulmonary disease in a Chinese population. *Acta Pharmacol Sin,* 37: 390-397.

[197] Hardin, M., Cho, M. H., McDonald, M. L., Wan, E., Lomas, D. A., Coxson, H. O., MacNee, W., Vestbo, J., Yates, J. C., Agusti, A., Calverley, P. M. A., Celli, B., Crim, C., Rennard, S., Wouters, E., Bakke, P., Bhatt, S. P., Kim, V., Ramsdell, J., Regan, E. A., Make, B. J., Hokanson, J. E., Crapo, J. D., Beaty, T. H., Hersh, C. P., Investigator, E., and Investigator, C. (2016). A genome-wide analysis of the response to inhaled beta(2)-agonists in chronic obstructive pulmonary disease. *Pharmacogenomics J,* 16: 326-335.

[198] Kim, H. J., Min, J. Y., Min, K. B., Seo, Y. S., Sung, J., Yun, J. M., Kwon, H., Cho, B., Park, J. H., and Kim, J. I. (2017). CDH13 gene-by-PM10 interaction effect on lung function decline in Korean men. *Chemosphere,* 168: 583-589.

[199] Takeuchi, T., Misaki, A., Fujita, J., Sonobe, H., and Ohtsuki, Y. (2001). T-cadherin (CDH13, H-cadherin) expression downregulated surfactant protein D in bronchioloalveolar cells. *Virchows Archiv-an International Journal of Pathology,* 438: 370-375.

[200] Sood, A., and Shore, S. A. (2013). Adiponectin, Leptin, and Resistin in Asthma: Basic Mechanisms through Population Studies. *J Allergy (Cairo),* 2013: 785835.

[201] Williams, A. S., Kasahara, D. I., Verbout, N. G., Fedulov, A. V., Zhu, M., Si, H., Wurmbrand, A. P., Hug, C., Ranscht, B., and Shore, S. A. (2012). Role of the adiponectin binding protein, T-cadherin (Cdh13), in allergic airways responses in mice. *PloS one,* 7: e41088.

[202] Kasahara, D. I., Williams, A. S., Benedito, L. A., Ranscht, B., Kobzik, L., Hug, C., and Shore, S. A. (2013). Role of the adiponectin binding protein, T-cadherin (cdh13), in pulmonary responses to subacute ozone. *PloS one,* 8: e65829.

[203] Yu, J., Ni, M., Xu, J., Zhang, H. Y., Gao, B. M., Gu, J. R., Chen, J. G., Zhang, L. S., Wu, M. C., Zhen, S. S., and Zhu, J. D. (2002). Methylation profiling of twenty promoter-CpG islands of genes

which may contribute to hepatocellular carcinogenesis. *Bmc Cancer,* 2.

[204] Khetani, S. R., Szulgit, G., Del Rio, J. A., Barlow, C., and Bhatia, S. N. (2004). Exploring interactions between rat hepatocytes and nonparenchymal cells using gene expression profiling. *Hepatology,* 40: 545-554.

[205] Khetani, S. R., Chen, A. A., Ranscht, B., and Bhatia, S. N. (2008). T-cadherin modulates hepatocyte functions in vitro. *Faseb J,* 22: 3768-3775.

[206] Arnemann, J., Sultani, O., Hasgun, D., and Coerdt, W. (2006). T-/H-cadherin (CDH13): a new marker for differentiating podocytes. *Virchows Archiv,* 448: 160-164.

[207] Hadjiargyrou, M., Lombardo, F., Zhao, S. C., Ahrens, W., Joo, J., Ahn, H., Jurman, M., White, D. W., and Rubin, C. T. (2002). Transcriptional profiling of bone regeneration - Insight into the molecular complexity of wound repair. *Journal of Biological Chemistry,* 277: 30177-30182.

[208] Wang, Q., Cai, J., Wang, J., Xiong, C., Yan, L., Zhang, Z., Fang, Y., and Zhao, J. (2014). Down-regulation of adiponectin receptors in osteoarthritic chondrocytes. *Cell biochemistry and biophysics,* 70: 491-497.

[209] Listyo, A., Brand, Y., Setz, C., Radojevic, V., Resink, T., Levano, S., and Bodmer, D. (2011). T-Cadherin in the Mammalian Cochlea. *Laryngoscope,* 121: 2228-2233.

[210] Simonneau, L., Gallego, M., and Pujol, R. (2003). Comparative expression patterns of T-, N-, E-cadherins, beta-catenin, and polysialic acid neural cell adhesion molecule in rat cochlea during development: Implications for the nature of Kolliker's organ. *J Comp Neurol,* 459: 113-126.

[211] Schepeler, T., Page, M. E., and Jensen, K. B. (2014). Heterogeneity and plasticity of epidermal stem cells. *Development,* 141: 2559-2567.

[212] Buechner, S. A., Philippova, M., Erne, P., Mathys, T., and Resink, T. J. (2009). High T-cadherin expression is a feature of basal cell carcinoma. *Brit J Dermatol,* 161: 199-202.

[213] Zhou, S. X., Matsuyoshi, N., Liang, S. B., Takeuchi, T., Ohtsuki, Y., and Miyachi, Y. (2002). Expression of T-cadherin in basal keratinocytes of skin. *J Invest Dermatol,* 118: 1080-1084.

[214] Buechner, S., Erne, P., and Resink, T. J. (2016). T-Cadherin Expression in the Epidermis and Adnexal Structures of Normal Skin. *Dermatopathology,* 3: 68-78.

[215] Kuphal, S., Martyn, A. C., Pedley, J., Crowther, L. M., Bonazzi, V. F., Parsons, P. G., Bosserhoff, A. K., Hayward, N. K., and Boyle, G. M. (2009). H-Cadherin expression reduces invasion of malignant melanoma. *Pigm Cell Melanoma R,* 22: 296-306.

[216] Won, C. H., Yoo, H. G., Park, K. Y., Shin, S. H., Park, W. S., Park, P. J., Chung, J. H., Kwon, O. S., and Kim, K. H. (2012). Hair Growth-Promoting Effects of Adiponectin In Vitro. *J Invest Dermatol,* 132: 2849-2851.

[217] Takeuchi, T., Liang, S. B., and Ohtsuki, Y. (2002). Downregulation of expression of a novel cadherin molecule, T-cadherin, in basal cell carcinoma of the skin. *Mol Carcinogen,* 35: 173-179.

[218] Li, L., Jiang, M., Feng, Q., Kiviat, N. B., Stern, J. E., Hawes, S., Cherne, S., and Lu, H. (2015). Aberrant Methylation Changes Detected in Cutaneous Squamous Cell Carcinoma of Immunocompetent Individuals. *Cell biochemistry and biophysics,* 72: 599-604.

[219] Pfaff, D., Philippova, M., Buechner, S. A., Maslova, K., Mathys, T., Erne, P., and Resink, T. J. (2010). T-cadherin loss induces an invasive phenotype in human keratinocytes and squamous cell carcinoma (SCC) cells *in vitro* and is associated with malignant transformation of cutaneous SCC in vivo. *Brit J Dermatol,* 163: 353-363.

[220] Takeuchi, T., Liang, S. B., Matsuyoshi, N., Zhou, S. X., Miyachi, Y., Sonobe, H., and Ohtsuki, Y. (2002). Loss of T-cadherin

(CDH13, H-cadherin) expression in cutaneous squamous cell carcinoma. *Lab Invest,* 82: 1023-1029.

[221] Buechner, S. A., and Resink, T. J. (2019). T-Cadherin Expression in Actinic Keratosis Transforming to Invasive Squamous Cell Carcinoma. *Dermatopathology,* 6: 12-19.

[222] Mukoyama, Y., Zhou, S. X., Miyachi, Y., and Matsuyoshi, N. (2005). T-cadherin negatively regulates the proliferation of cutaneous squamous carcinoma cells. *J Invest Dermatol,* 124: 833-838.

[223] Pfaff, D., Philippova, M., Kyriakakis, E., Maslova, K., Rupp, K., Buechner, S. A., Iezzi, G., Spagnoli, G. C., Erne, P., and Resink, T. J. (2011). Paradoxical effects of T-cadherin on squamous cell carcinoma: up- and down-regulation increase xenograft growth by distinct mechanisms. *J Pathol,* 225: 512-524.

[224] Philippova, M., Pfaff, D., Kyriakakis, E., Buechner, S. A., Iezzi, G., Spagnoli, G. C., Schoenenberger, A. W., Erne, P., and Resink, T. J. (2013). T-cadherin loss promotes experimental metastasis of squamous cell carcinoma. *Eur J Cancer,* 49: 2048-2058.

[225] Kyriakakis, E., Maslova, K., Philippova, M., Pfaff, D., Joshi, M. B., Buechner, S. A., Erne, P., and Resink, T. J. (2012). T-Cadherin Is an Auxiliary Negative Regulator of EGFR Pathway Activity in Cutaneous Squamous Cell Carcinoma: Impact on Cell Motility. *J Invest Dermatol,* 132: 2275-2285.

[226] Kyriakakis, E., Maslova, K., Frachet, A., Ferri, N., Contini, A., Pfaff, D., Erne, P., Resink, T. J., and Philippova, M. (2013). Crosstalk between EGFR and T-cadherin: EGFR activation promotes T-cadherin localization to intercellular contacts. *Cell Signal,* 25: 1044-1053.

[227] Mukoyama, Y., Utani, A., Matsui, S., Zhou, S. X., Miyachi, Y., and Matsuyoshi, N. (2007). T-cadherin enhances cell-matrix adhesiveness by regulating beta 1 integrin trafficking in cutaneous squamous carcinoma cells. *Genes Cells,* 12: 787-796.

[228] Duan, X. S., Lu, J., Ge, Z. H., Xing, E. H., Lu, H. T., and Sun, L. X. (2013). Effects of T-cadherin expression on B16F10 melanoma cells. *Oncology letters,* 5: 1205-1210.

[229] Bosserhoff, A. K., Ellmann, L., Quast, A. S., Eberle, J., Boyle, G. M., and Kuphal, S. (2014). Loss of T-Cadherin (CDH-13) Regulates AKT Signaling and Desensitizes Cells to Apoptosis in Melanoma. *Mol Carcinogen,* 53: 635-647.

[230] Ellmann, L., Joshi, M. B., Resink, T. J., Bosserhoff, A. K., and Kuphal, S. (2012). BRN2 is a transcriptional repressor of CDH13 (T-cadherin) in melanoma cells. *Lab Invest,* 92: 1788-1800.

[231] Rubina, K. A., Surkova, E. I., Semina, E. V., Sysoeva, V. Y., Kalinina, N. I., Poliakov, A. A., Treshalina, H. M., and Tkachuk, V. A. (2015). T-Cadherin Expression in Melanoma Cells Stimulates Stromal Cell Recruitment and Invasion by Regulating the Expression of Chemokines, Integrins and Adhesion Molecules. *Cancers,* 7: 1349-1370.

[232] Zhou, S., Matsuyoshi, N., Takeuchi, T., Ohtsuki, Y., and Miyachi, Y. (2003). Reciprocal altered expression of T-cadherin and P-cadherin in psoriasis vulgaris. *The British journal of dermatology,* 149: 268-273.

[233] Lee, S. W. (1996). H-cadherin, a novel cadherin with growth inhibitory functions and diminished expression in human breast cancer. *Nat Med,* 2: 776-782.

[234] Toyooka, K. O., Toyooka, S., Virmani, A. K., Sathyanarayana, U. G., Euhus, D. M., Gilcrease, M., Minna, J. D., and Gazdar, A. F. (2001). Loss of expression and aberrant methylation of the CDH13 (H-cadherin) gene in breast and lung carcinomas. *Cancer research,* 61: 4556-4560.

[235] Hafez, M. M., Al-Shabanah, O. A., Al-Rejaie, S. S., Al-Harbi, N. O., Hassan, Z. K., Alsheikh, A., Al Theyab, A. I., Aldelemy, M. L., and Sayed-Ahmed, M. M. (2015). Increased hypermethylation of glutathione S-transferase P1, DNA-binding protein inhibitor, death associated protein kinase and paired box protein-5 genes in triple-

negative breast cancer Saudi females. *Asian Pacific journal of cancer prevention : APJCP,* 16: 541-549.

[236] Miki, Y., Katagiri, T., and Nakamura, Y. (1997). Infrequent mutation of the H-cadherin gene on chromosome 16q24 in human breast cancers. *Japanese journal of cancer research: Gann,* 88: 701-704.

[237] Riener, M. O., Nikolopoulos, E., Herr, A., Johannes, P., Hausmann, M., Wiech, T., Orlowska-Volk, M., Lassmann, S., Walch, A., and Werner, M. (2008). Microarray comparative genomic hybridization analysis of tubular breast carcinoma shows recurrent toss of the CDH13 locus on 16q. *Hum Pathol,* 39: 1621-1629.

[238] Xu, J., Shetty, P. B., Feng, W. W., Chenault, C., Bast, R. C., Issa, J. P. J., Hilsenbeck, S. G., and Yu, Y. H. (2012). Methylation of HIN-1, RASSF1A, RIL and CDH13 in breast cancer is associated with clinical characteristics, but only RASSF1A methylation is associated with outcome. *Bmc Cancer,* 12.

[239] Feng, W. W., Orlandi, R., Zhao, N. Q., Carcangiu, M. L., Tagliabue, E., Xu, J., Bast, R. C., and Yu, Y. (2010). Tumor suppressor genes are frequently methylated in lymph node metastases of breast cancers. *Bmc Cancer,* 10.

[240] Takeuchi, T., Misaki, A., Chen, B. K., and Ohtsuki, Y. (1999). H-cadherin expression in breast cancer. *Histopathology,* 35: 87-88.

[241] Lewis, C. M., Cler, L. R., Bu, D. W., Zochbauer-Muller, S., Milchgrub, S., Naftalis, E. Z., Leitch, A. M., Minna, J. D., and Euhus, D. M. (2005). Promoter hypermethylation in benign breast epithelium in relation to predicted breast cancer risk. *Clin Cancer Res,* 11: 166-172.

[242] Verschuur-Maes, A. H. J., de Bruin, P. C., and Diest, P. J. (2012). Epigenetic progression of columnar cell lesions of the breast to invasive breast cancer. *Breast cancer research and treatment,* 136: 705-715.

[243] Kong, D. D., Yang, J., Li, L., Wang, W., Chen, Y. N., Wang, S. B., and Zhou, Y. Z. (2015). T-cadherin association with clinicopathological features and prognosis in axillary lymph node-

positive breast cancer. *Breast cancer research and treatment,* 150: 119-126.

[244] Kong, D. D., Wang, M. H., Yang, J., Li, L., Wang, W., Wang, S. B., and Zhou, Y. Z. (2017). T-cadherin is associated with prognosis in triple-negative breast cancer. *Oncology letters,* 14: 2975-2981.

[245] Kong, D., Wang, M. H., Yang, J., and Li, L. (2017). Association of T-cadherin levels with the response to neoadjuvant chemotherapy in locally advanced breast cancer. *Oncotarget,* 8: 13747-13753.

[246] Wang, S. P., Dorsey, T. H., Terunuma, A., Kittles, R. A., Ambs, S., and Kwabi-Addo, B. (2012). Relationship between Tumor DNA Methylation Status and Patient Characteristics in African-American and European-American Women with Breast Cancer. *PloS one,* 7.

[247] Jeong, Y. J., Jeong, H. Y., Bong, J. G., Park, S. H., and Oh, H. K. (2013). Low methylation levels of the SFRP1 gene are associated with the basal-like subtype of breast cancer. *Oncol Rep,* 29: 1946-1954.

[248] Fiegl, H., Millinger, S., Goebel, G., Muller-Holzner, E., Marth, C., Laird, P. W., and Widschwendter, M. (2006). Breast cancer DNA methylation profiles in cancer cells and tumor stroma: Association with HER-2/neu status in primary breast cancer. *Cancer research,* 66: 29-33.

[249] Celebiler Cavusoglu, A., Kilic, Y., Saydam, S., Canda, T., Baskan, Z., Sevinc, A. I., and Sakizli, M. (2009). Predicting invasive phenotype with CDH1, CDH13, CD44, and TIMP3 gene expression in primary breast cancer. *Cancer science,* 100: 2341-2345.

[250] Denoyelle, C., Albanese, P., Uzan, G., Hong, L., Vannier, J. P., Soria, J., and Soria, C. (2003). Molecular mechanism of the anti-cancer activity of cerivastatin, an inhibitor of HMG-CoA reductase, on aggressive human breast cancer cells. *Cell Signal,* 15: 327-338.

[251] Lee, S. W., Reimer, C. L., Campbell, D. B., Cheresh, P., Duda, R. B., and Kocher, O. (1998). H-cadherin expression inhibits in vitro invasiveness and tumor formation in vivo. *Carcinogenesis,* 19: 1157-1159.

[252] Wang, X. D., Wang, B. E., Soriano, R., Zha, J., Zhang, Z., Modrusan, Z., Cunha, G. R., and Gao, W. Q. (2007). Expression profiling of the mouse prostate after castration and hormone replacement: implication of H-cadherin in prostate tumorigenesis. *Differentiation; research in biological diversity,* 75: 219-234.

[253] Thomas, G., Jacobs, K. B., Yeager, M., Kraft, P., Wacholder, S., Orr, N., Yu, K., Chatterjee, N., Welch, R., Hutchinson, A., Crenshaw, A., Cancel-Tassin, G., Staats, B. J., Wang, Z., Gonzalez-Bosquet, J., Fang, J., Deng, X., Berndt, S. I., Calle, E. E., Feigelson, H. S., Thun, M. J., Rodriguez, C., Albanes, D., Virtamo, J., Weinstein, S., Schumacher, F. R., Giovannucci, E., Willett, W. C., Cussenot, O., Valeri, A., Andriole, G. L., Crawford, E. D., Tucker, M., Gerhard, D. S., Fraumeni, J. F., Hoover, R., Hayes, R. B., Hunter, D. J., and Chanock, S. J. (2008). Multiple loci identified in a genome-wide association study of prostate cancer. *Nat Genet,* 40: 310-315.

[254] Maruyama, R., Toyooka, S., Toyooka, K. O., Virmani, A. K., Zochbauer-Muller, S., Farinas, A. J., Minna, J. D., McConnell, J., Frenkel, E. P., and Gazdar, A. F. (2002). Aberrant promoter methylation profile of prostate cancers and its relationship to clinicopathological features. *Clin Cancer Res,* 8: 514-519.

[255] Dasen, B., Vlajnic, T., Mengus, C., Ruiz, C., Bubendorf, L., Spagnoli, G., Wyler, S., Erne, P., Resink, T. J., and Philippova, M. (2017). T-cadherin in prostate cancer: relationship with cancer progression, differentiation and drug resistance. *J Pathol Clin Res,* 3: 44-57.

[256] Maslova, K., Kyriakakis, E., Pfaff, D., Frachet, A., Frismantiene, A., Bubendorf, L., Ruiz, C., Vlajnic, T., Erne, P., Resink, T. J., and Philippova, M. (2015). EGFR and IGF-1R in regulation of prostate cancer cell phenotype and polarity: opposing functions and modulation by T-cadherin. *Faseb J,* 29: 494-507.

[257] Wu, W. F., Maneix, L., Insunza, J., Nalvarte, I., Antonson, P., Kere, J., Yu, N. Y. L., Tohonen, V., Katayama, S., Einarsdottir, E., Krjutskov, K., Dai, Y. B., Huang, B., Su, W., Warner, M., and

Gustafsson, J. A. (2017). Estrogen receptor beta, a regulator of androgen receptor signaling in the mouse ventral prostate. *P Natl Acad Sci USA,* 114: E3816-E3822.

[258] Zhang, D. X., Park, D., Zhong, Y., Lu, Y., Rycaj, K., Gong, S., Chen, X., Liu, X., Chao, H. P., Whitney, P., Calhoun-Davis, T., Takata, Y., Shen, J. J., Iyer, V. R., and Tang, D. G. (2016). Stem cell and neurogenic gene-expression profiles link prostate basal cells to aggressive prostate cancer. *Nat Commun,* 7.

[259] Smith, B. A., Sokolov, A., Uzunangelov, V., Baertsch, R., Newton, Y., Graim, K., Mathis, C., Cheng, D. H., Stuart, J. M., and Witte, O. N. (2015). A basal stem cell signature identifies aggressive prostate cancer phenotypes. *P Natl Acad Sci USA,* 112: E6544-E6552.

[260] Wang, L., Lin, Y. L., Li, B., Wang, Y. Z., Li, W. P., and Ma, J. G. (2014). Aberrant promoter methylation of the cadherin 13 gene in serum and its relationship with clinicopathological features of prostate cancer. *The Journal of international medical research,* 42: 1085-1092.

[261] Moses-Fynn, E., Tang, W., Beyene, D., Apprey, V., Copeland, R., Kanaan, Y., and Kwabi-Addo, B. (2018). Correlating blood-based DNA methylation markers and prostate cancer risk in African-American men. *PloS one,* 13.

[262] Sato, M., Mori, Y., Sakurada, A., Fujimura, S., and Horii, A. (1998). The H-cadherin (CDH13) gene is inactivated in human lung cancer. *Human genetics,* 103: 96-101.

[263] Kim, D. S., Kim, M. J., Lee, J. Y., Kim, Y. Z., Kim, E. J., and Park, J. Y. (2007). Aberrant methylation of E-cadherin and H-cadherin genes in nonsmall cell lung cancer and its relation to clinicopathologic features. *Cancer,* 110: 2785-2792.

[264] Brock, M. V., Hooker, C. M., Ota-Machida, E., Han, Y., Guo, M. Z., Ames, S., Glockner, S., Piantadosi, S., Gabrielson, E., Pridham, G., Pelosky, K., Belinsky, S. A., Yang, S. C., Baylin, S. B., and Herman, J. G. (2008). DNA methylation markers and early recurrence in stage I lung cancer. *New Engl J Med,* 358: 1118-1128.

[265] Zhong, Y., Delgado, Y., Gomez, J., Lee, S. W., and Perez-Soler, R. (2001). Loss of H-cadherin protein expression in human non-small cell lung cancer is associated with tumorigenicity. *Clin Cancer Res,* 7: 1683-1687.

[266] Zhong, Y. H., Peng, H., Cheng, H. Z., and Wang, P. (2015). Quantitative assessment of the diagnostic role of CDH13 promoter methylation in lung cancer. *Asian Pacific journal of cancer prevention : APJCP,* 16: 1139-1143.

[267] Ulivi, P., Zoli, W., Calistri, D., Fabbri, F., Tesei, A., Rosetti, M., Mengozzi, M., and Amadori, D. (2006). p16INK4A and CDH13 hypermethylation in tumor and serum of non-small cell lung cancer patients. *Journal of cellular physiology,* 206: 611-615.

[268] Kim, J. S., Han, J. H., Shim, Y. M., Park, J., and Kim, D. H. (2005). Aberrant methylation of H-Cadherin (CDH13) promoter is associated with tumor progression in primary nonsmall cell lung carcinoma. *Cancer,* 104: 1825-1833.

[269] Kontic, M., Stojsic, J., Jovanovic, D., Bunjevacki, V., Ognjanovic, S., Kuriger, J., Puumala, S., and Nelson, H. H. (2012). Aberrant Promoter Methylation of CDH13 and MGMT Genes is Associated With Clinicopathologic Characteristics of Primary Non-Small-Cell Lung Carcinoma. *Clin Lung Cancer,* 13: 297-303.

[270] Saito, K., Kawakami, K., Matsumoto, I., Oda, M., Watanabe, G., and Minamoto, T. (2010). Long Interspersed Nuclear Element 1 Hypomethylation Is a Marker of Poor Prognosis in Stage IA Non-Small Cell Lung Cancer. *Clin Cancer Res,* 16: 2418-2426.

[271] Kubo, T., Yamamoto, H., Ichimura, K., Jida, M., Hayashi, T., Otani, H., Tsukuda, K., Sano, Y., Kiura, K., and Toyooka, S. (2009). DNA methylation in small lung adenocarcinoma with bronchioloalveolar carcinoma components. *Lung Cancer,* 65: 328-332.

[272] Wang, Y., Zhang, D. D., Zheng, W. L., Luo, J. F., Bai, Y. F., and Lu, Z. H. (2008). Multiple gene methylation of nonsmall cell lung cancers evaluated with 3-dimensional microarray. *Cancer,* 112: 1325-1336.

[273] Tsou, J. A., Galler, J. S., Siegmund, K. D., Laird, P. W., Turla, S., Cozen, W., Hagen, J. A., Koss, M. N., and Laird-Offringa, I. A. (2007). Identification of a panel of sensitive and specific DNA methylation markers for lung adenocarcinoma. *Mol Cancer,* 6.

[274] Hanabata, T., Tsukuda, K., Toyooka, S., Yano, M., Aoe, M., Nagahiro, I., Sano, Y., Date, H., and Shimizu, N. (2004). DNA methylation of multiple genes and clinicopathological relationship of non-small cell lung cancers. *Oncol Rep,* 12: 177-180.

[275] Feng, Q. H., Hawes, S. E., Stern, J. E., Wiens, L., Lu, H., Dong, Z. M., Jordanj, C. D., Kiviatl, N. B., and Vesselle, H. (2008). DNA methylation in tumor and matched normal tissues from non-small cell lung cancer patients. *Cancer Epidem Biomar,* 17: 645-654.

[276] Kim, H. J., Kwon, Y. M., Kim, J. S., Han, J. H., Shim, Y. M., Park, J. B., and Kim, D. H. (2006). Elevated mRNA levels of DNA methyltransferase-1 as an independent prognostic factor in primary nonsmall cell lung cancer. *Cancer,* 107: 1042-1049.

[277] Kim, J. S., Kim, J. W., Han, J. H., Shim, Y. M., Park, J., and Kim, D. H. (2006). Cohypermethylation of p16 and FHIT promoters as a prognostic factor of recurrence in surgically resected stage I non-small cell lung cancer. *Cancer research,* 66: 4049-4054.

[278] Maruyama, R., Sugio, K., Yoshino, L., Maehara, Y., and Gazdar, A. F. (2004). Hypermethylation of FHIT as a prognostic marker in nonsmall cell lung carcinoma. *Cancer,* 100: 1472-1477.

[279] Toyooka, S., Toyooka, K. O., Maruyama, R., Virmani, A. K., Girard, L., Miyajima, K., Harada, K., Ariyoshi, Y., Takahashi, T., Sugio, K., Brambilla, E., Gilcrease, M., Minna, J. D., and Gazdar, A. F. (2001). DNA methylation profiles of lung tumors. *Mol Cancer Ther,* 1: 61-67.

[280] Zhai, X., and Li, S. J. (2014). Methylation of RASSF1A and CDH13 Genes in Individualized Chemotherapy for Patients with Non-small Cell Lung Cancer. *Asian Pac J Cancer P,* 15: 4925-4928.

[281] Jin, M. J., Kawakami, K., Fukui, Y., Tsukioka, S., Oda, M., Watanabe, G., Takechi, T., Oka, T., and Minamoto, T. (2009). Different histological types of non-small cell lung cancer have

distinct folate and DNA methylation levels. *Cancer science,* 100: 2325-2330.

[282] Li, Y., Li, C., Ma, Q., Zhang, Y., Yao, Y., Liu, S., Zhang, X., Hong, C., Tan, F., Shi, L., and Yao, Y. (2018). Genetic variation in CDH13 gene was associated with non-small cell lung cancer (NSCLC): A population-based case-control study. *Oncotarget,* 9: 881-891.

[283] Wang, Z., Wang, B., Guo, H., Shi, G., and Hong, X. (2015). Clinicopathological significance and potential drug target of T-cadherin in NSCLC. *Drug design, development and therapy,* 9: 207-216.

[284] Toyooka, S., Matsuo, K., and Gazdar, A. F. (2008). DNA methylation in lung cancer. *New Engl J Med,* 358: 2513-2513.

[285] Huang, T., Li, J., Zhang, C., Hong, Q., Jiang, D., Ye, M., and Duan, S. (2016). Distinguishing Lung Adenocarcinoma from Lung Squamous Cell Carcinoma by Two Hypomethylated and Three Hypermethylated Genes: A Meta-Analysis. *PloS one,* 11: e0149088.

[286] Hsu, H. S., Chen, T. P., Hung, C. H., Wen, C. K., Lin, R. K., Lee, H. C., and Wang, Y. C. (2007). Characterization of a multiple epigenetic marker panel for lung cancer detection and risk assessment in plasma. *Cancer,* 110: 2019-2026.

[287] Toyooka, S., Suzuki, M., Maruyama, R., Toyooka, K. O., Tsukuda, K., Fukuyama, Y., Iizasa, T., Aoe, M., Date, H., Fujisawa, T., Shimizu, N., and Gazdar, A. F. (2004). The relationship between aberrant methylation and survival in non-small-cell lung cancers. *Brit J Cancer,* 91: 771-774.

[288] Yanagawa, N., Tamura, G., Oizumi, H., Kanauchi, N., Endoh, M., Sadahiro, M., and Motoyama, T. (2007). Promoter hypermethylation of RASSF1A and RUNX3 genes as an independent prognostic prediction marker in surgically resected non-small cell lung cancers. *Lung Cancer,* 58: 131-138.

[289] Vuillemenot, B. R., Hutt, J. A., and Belinsky, S. A. (2006). Gene promoter hypermethylation in mouse lung tumors. *Mol Cancer Res,* 4: 267-273.

[290] Blanco, D., Vicent, S., Fraga, M. F., Fernandez-Garcia, I., Freire, J., Lujambio, A., Esteller, M., Ortiz-de-Solorzano, C., Pio, R., Lecanda, F., and Montuenga, L. M. (2007). Molecular analysis of a multistep lung cancer model induced by chronic inflammation reveals epigenetic regulation of p16 and activation of the DNA damage response pathway. *Neoplasia,* 9: 840-852.

[291] Kim, H., Kwon, Y. M., Kim, J. S., Lee, H., Park, J. H., Shim, Y. M., Han, J., Park, J., and Kim, D. H. (2004). Tumor-specific methylation in bronchial lavage for the early detection of non-small-cell lung cancer. *J Clin Oncol,* 22: 2363-2370.

[292] Liu, J. F., Li, Y. S., Drew, P. A., and Zhang, C. (2016). The effect of celecoxib on DNA methylation of CDH13, TFPI2, and FSTL1 in squamous cell carcinoma of the esophagus in vivo. *Anti-cancer drugs,* 27: 848-853.

[293] Hibi, K., Kodera, Y., Ito, K., Akiyama, S., and Nakao, A. (2004). Methylation pattern of CDH13 gene in digestive tract cancers. *Brit J Cancer,* 91: 1139-1142.

[294] Jin, Z., Cheng, Y. L., Olaru, A., Kan, T., Yang, J., Paun, B., Ito, T., Hamilton, J. P., David, S., Agarwal, R., Selaru, F. M., Sato, F., Abraham, J. M., Beer, D. G., Mori, Y., Shimada, Y., and Meltzer, S. J. (2008). Promoter hypermethylation of CDH13 is a common, early event in human esophageal adenocarcinogenesis and correlates with clinical risk factors. *Int J Cancer,* 123: 2331-2336.

[295] Guo, Q., Wang, H. B., Li, Y. H., Li, H. F., Li, T. T., Zhang, W. X., Xiang, S. S., and Sun, Z. Q. (2016). Correlations of Promoter Methylation in WIF-1, RASSF1A, and CDH13 Genes with the Risk and Prognosis of Esophageal Cancer. *Medical science monitor: international medical journal of experimental and clinical research,* 22: 2816-2824.

[296] Fukuoka, T., Hibi, K., and Nakao, A. (2006). Aberrant methylation is frequently observed in advanced esophageal squamous cell carcinoma. *Anticancer Res,* 26: 3333-3335.

[297] Mori, Y., Matsunaga, M., Abe, T., Fukushige, S., Miura, K., Sunamura, M., Shiiba, K., Sato, M., Nukiwa, T., and Horii, A.

(1999). Chromosome band 16q24 is frequently deleted in human gastric cancer. *Brit J Cancer,* 80: 556-562.
[298] Tang, Y. P., Dai, Y. H., and Huo, J. R. (2012). Decreased Expression of T-Cadherin is Associated with Gastric Cancer Prognosis. *Hepato-Gastroenterol,* 59: 1294-1298.
[299] Wei, B., Shi, H., Lu, X., Shi, A., Cheng, Y., and Dong, L. (2015). Association between the expression of T-cadherin and vascular endothelial growth factor and the prognosis of patients with gastric cancer. *Molecular medicine reports,* 12: 2075-2081.
[300] Lin, J., Chen, Z., Huang, Z., Chen, F., Ye, Z., Lin, S., and Wang, W. (2019). Effect of T-cadherin on the AKT/mTOR signaling pathway, gastric cancer cell cycle, migration and invasion, and its association with patient survival rate. *Experimental and therapeutic medicine,* 17: 3607-3613.
[301] Lin, J., Chen, Z., Huang, Z., Chen, F., Ye, Z., Lin, S., and Wang, W. (2017). Upregulation of T-cadherin suppresses cell proliferation, migration and invasion of gastric cancer in vitro. *Experimental and therapeutic medicine,* 14: 4194-4200.
[302] Berkhout, M., Nagtegaal, I. D., Cornelissen, S. J. B., Dekkers, M. M. G., de Molengraft, F. J. J. M. V., Peters, W. H. M., Nagengast, F. M., van Krieken, J. H. J. M., and Jeuken, J. W. M. (2007). Chromosomal and methylation alterations in sporadic and familial adenomatous polyposis-related duodenal carcinomas. *Modern Pathol,* 20: 1253-1262.
[303] Leong, K. J., Wei, W., Tannahill, L. A., Caldwell, G. M., Jones, C. E., Morton, D. G., Matthews, G. M., and Bach, S. P. (2011). Methylation profiling of rectal cancer identifies novel markers of early-stage disease. *Brit J Surg,* 98: 724-734.
[304] Joensuu, E. I., Abdel-Rahman, W. M., Ollikainen, M., Ruosaari, S., Knuutila, S., and Peltomaki, P. (2008). Epigenetic signatures of familial cancer are characteristic of tumor type and family category. *Cancer research,* 68: 4597-4605.

[305] Luo, L. P., Chen, W. D., and Pretlow, T. P. (2005). CpG island methylation in aberrant crypt foci and cancers from the same patients. *Int J Cancer,* 115: 747-751.

[306] Toyooka, S., Toyooka, K. O., Harada, K., Miyajima, K., Makarla, P., Sathyanarayana, U. G., Yin, J., Sato, F., Shivapurkar, N., Meltzer, S. J., and Gazdar, A. F. (2002). Aberrant methylation of the CDH13 (H-cadherin) promoter region in colorectal cancers and adenomas. *Cancer research,* 62: 3382-3386.

[307] Duan, B. S., Xie, L. F., and Wang, Y. (2017). Aberrant Methylation of T-cadherin Can Be a Diagnostic Biomarker for Colorectal Cancer. *Cancer genomics & proteomics,* 14: 277-284.

[308] Xu, X. L., Yu, J., Zhang, H. Y., Sun, M. H., Gu, J., Du, X., Shi, D. R., Wang, P., Yang, Z. H., and Zhu, J. D. (2004). Methylation profile of the promoter CpG islands of 31 genes that may contribute to colorectal carcinogenesis. *World journal of gastroenterology,* 10: 3441-3454.

[309] Scarpa, M., Scarpa, M., Castagliuolo, I., Erroi, F., Kotsafti, A., Basato, S., Brun, P., D'Inca, R., Rugge, M., Angriman, I., and Castoro, C. (2016). Aberrant gene methylation in non-neoplastic mucosa as a predictive marker of ulcerative colitis-associated CRC. *Oncotarget,* 7: 10322-10331.

[310] Ye, M., Huang, T., Li, J. Y., Zhou, C. C., Yang, P., Ni, C., and Chen, S. (2017). Role of CDH13 promoter methylation in the carcinogenesis, progression, and prognosis of colorectal cancer A systematic meta-analysis under PRISMA guidelines. *Medicine,* 96.

[311] Ren, J. Z., and Huo, J. R. (2012). Correlation between T-cadherin gene expression and aberrant methylation of T-cadherin promoter in human colon carcinoma cells. *Med Oncol,* 29: 915-918.

[312] Hibi, K., Nakayama, H., Kodera, Y., Ito, K., Akiyama, S., and Nakao, A. (2004). CDH13 promoter region is specifically methylated in poorly differentiated colorectal cancer. *Brit J Cancer,* 90: 1030-1033.

[313] Wang, Z., Yuan, X., Jiao, N., Zhu, H., Zhang, Y., and Tong, J. (2012). CDH13 and FLBN3 gene methylation are associated with

poor prognosis in colorectal cancer. *Pathology oncology research : POR,* 18: 263-270.

[314] Hibi, K., Kodera, Y., Ito, K., Akiyama, S., and Nakao, A. (2005). Aberrant methylation of HLTF, SOCS-1, and CDH13 genes is shown in colorectal cancers without lymph node metastasis. *Dis Colon Rectum,* 48: 1282-1286.

[315] Hibi, K., and Nakao, A. (2006). Lymph node metastasis is infrequent in patients with highly-methylated colorectal cancer. *Anticancer Res,* 26: 55-58.

[316] Katira, A., and Tan, P. H. (2016). Evolving role of adiponectin in cancer-controversies and update. *Cancer Biol Med,* 13: 101-119.

[317] Tae, C. H., Kim, S. E., Jung, S. A., Joo, Y. H., Shim, K. N., Jung, H. K., Kim, T. H., Cho, M. S., Kim, K. H., and Kim, J. S. (2014). Involvement of adiponectin in early stage of colorectal carcinogenesis. *Bmc Cancer,* 14.

[318] Park, J., Kim, I., Jung, K. J., Kim, S., Jee, S. H., and Yoon, S. K. (2015). Gene-gene interaction analysis identifies a new genetic risk factor for colorectal cancer. *J Biomed Sci,* 22.

[319] Zhu, C., Feng, X., Ye, G., and Huang, T. (2017). Meta-analysis of possible role of cadherin gene methylation in evolution and prognosis of hepatocellular carcinoma with a PRISMA guideline. *Medicine,* 96: e6650.

[320] Chan, D. W., Lee, J. M. F., Chan, P. C. Y., and Ng, I. O. L. (2008). Genetic and epigenetic inactivation of T-cadherin in human hepatocellular carcinoma cells. *Int J Cancer,* 123: 1043-1052.

[321] Yamada, S., Nomoto, S., Fujii, T., Takeda, S., Kanazumi, N., Sugimoto, H., and Nakao, A. (2007). Frequent promoter methylation of M-cadherin in hepatocellular carcinoma is associated with poor prognosis. *Anticancer Res,* 27: 2269-2274.

[322] Yan, Q., Zhang, Z. F., Chen, X. P., Gutmann, D. H., Xiong, M., Xiao, Z. Y., and Huang, Z. Y. (2008). Reduced T-cadherin expression and promoter methylation are associated with the development and progression of hepatocellular carcinoma. *International journal of oncology,* 32: 1057-1063.

[323] Riou, P., Saffroy, R., Comoy, J., Gross-Goupil, M., Thiery, J. P., Emile, J. F., Azoulay, D., Piatier-Tonneau, D., Lemoine, A., and Debuire, B. (2002). Investigation in liver tissues and cell lines of the transcription of 13 genes mapping to the 16q24 region that are frequently deleted in hepatocellular carcinoma. *Clin Cancer Res,* 8: 3178-3186.

[324] Riou, P., Saffroy, R., Chenailler, C., Franc, B., Gentile, C., Rubinstein, E., Resink, T., Debuire, B., Piatier-Tonneau, D., and Lemoine, A. (2006). Expression of T-cadherin in tumor cells influences invasive potential of human hepatocellular carcinoma. *Faseb J,* 20: 2291-2301.

[325] Sakai, M., Hibi, K., Koshikawa, K., Inoue, S., Takeda, S., Kaneko, T., and Nakao, A. (2004). Frequent promoter methylation and gene silencing of CDH13 in pancreatic cancer. *Cancer science,* 95: 588-591.

[326] Chen, F., Huang, T., Ren, Y., Wei, J., Lou, Z., Wang, X., Fan, X., Chen, Y., Weng, G., and Yao, X. (2016). Clinical significance of CDH13 promoter methylation as a biomarker for bladder cancer: a meta-analysis. *BMC urology,* 16: 52.

[327] Lin, Y. L., He, Z. K., Li, Z. G., and Guan, T. Y. (2013). Downregulation of CDH13 Expression Promotes Invasiveness of Bladder Transitional Cell Carcinoma. *Urol Int,* 90: 225-232.

[328] Lin, Y. L., Xie, P. G., and Ma, J. G. (2014). Aberrant methylation of CDH13 is a potential biomarker for predicting the recurrence and progression of non muscle invasive bladder cancer. *Medical science monitor : international medical journal of experimental and clinical research,* 20: 1572-1577.

[329] Lin, Y. L., Sun, G., Liu, X. Q., Li, W. P., and Ma, J. G. (2011). Clinical Significance of CDH13 Promoter Methylation in Serum Samples from Patients with Bladder Transitional Cell Carcinoma. *Journal of International Medical Research,* 39: 179-186.

[330] Lin, Y. L., Liu, X. Q., Li, W. P., Sun, G., and Zhang, C. T. (2012). Promoter methylation of H-cadherin is a potential biomarker in

patients with bladder transitional cell carcinoma. *Int Urol Nephrol,* 44: 111-117.

[331] Adachi, Y., Takeuchi, T., Nagayama, T., Ohtsuki, Y., and Furihata, M. (2009). Zeb1-mediated T-cadherin repression increases the invasive potential of gallbladder cancer. *FEBS letters,* 583: 430-436.

[332] Adachi, Y., Takeuchi, T., Nagayama, T., and Furihata, M. (2010). T-cadherin modulates tumor-associated molecules in gallbladder cancer cells. *Cancer investigation,* 28: 120-126.

[333] Xu, Y., Li, X., Wang, H., Xie, P., Yan, X., Bai, Y., and Zhang, T. (2016). Hypermethylation of CDH13, DKK3 and FOXL2 promoters and the expression of EZH2 in ovary granulosa cell tumors. *Molecular medicine reports,* 14: 2739-2745.

[334] Pawlik, P., Mostowska, A., Lianeri, M., Sajdak, S., Kedzia, H., and Jagodzinski, P. P. (2012). Folate and choline metabolism gene variants in relation to ovarian cancer risk in the Polish population. *Mol Biol Rep,* 39: 5553-5560.

[335] Bol, G. M., Suijkerbuijk, K. P. M., Bart, J., Vooijs, M., van der Wall, E., and van Diest, P. J. (2010). Methylation profiles of hereditary and sporadic ovarian cancer. *Histopathology,* 57: 363-370.

[336] Makarla, P. B., Saboorian, M. H., Ashfaq, R., Toyooka, K. O., Toyooka, S., Minna, J. D., Gazdar, A. F., and Schorge, J. O. (2005). Promoter hypermethylation profile of ovarian epithelial neoplasms. *Clin Cancer Res,* 11: 5365-5369.

[337] Kawakami, M., Staub, J., Cliby, W., Hartmann, L., Smith, D. I., and Shridhar, V. (1999). Involvement of H-cadherin (CDH13) on 16q in the region of frequent deletion in ovarian cancer. *International journal of oncology,* 15: 715-720.

[338] Sheng, Y., Wang, H., Liu, D., Zhang, C., Deng, Y., Yang, F., Zhang, T., and Zhang, C. (2016). Methylation of tumor suppressor gene CDH13 and SHP1 promoters and their epigenetic regulation by the UHRF1/PRMT5 complex in endometrial carcinoma. *Gynecologic oncology,* 140: 145-151.

[339] Suehiro, Y., Okada, T., Okada, T., Anno, K., Okayama, N., Ueno, K., Hiura, M., Nakamura, M., Kondo, T., Oga, A., Kawauchi, S., Hirabayashi, K., Numa, F., Ito, T., Saito, T., Sasaki, K., and Hinoda, Y. (2008). Aneuploidy predicts outcome in patients with endometrial carcinoma and is related to lack of CDH13 hypermethylation. *Clin Cancer Res,* 14: 3354-3361.

[340] Missaoui, N., Hmissa, S., Trabelsi, A., Traore, C., Mokni, M., Dante, R., and Frappart, L. (2011). Promoter hypermethylation of CDH13, DAPK1 and TWIST1 genes in precancerous and cancerous lesions of the uterine cervix. *Pathol Res Pract,* 207: 37-42.

[341] Widschwendter, A., Ivarsson, L., Blassnig, A., Muller, H. M., Fiegl, H., Wiedemarz, A., Muller-Holzner, E., Goebel, G., Marth, C., and Widschwendter, M. (2004). CDH1 and CDH13 methylation in serum is an independent prognostic marker in cervical cancer patients. *Int J Cancer,* 109: 163-166.

[342] Abudukadeer, A., Bakry, R., Goebel, G., Mutz-Dehbalaie, I., Widschwendter, A., Bonn, G. K., and Fiegl, H. (2012). Clinical Relevance of CDH1 and CDH13 DNA-Methylation in Serum of Cervical Cancer Patients. *International journal of molecular sciences,* 13: 8353-8363.

[343] Zhao, J., Yang, T., Ji, J., Li, C., Li, Z., and Li, L. (2018). Garcinol exerts anti-cancer effect in human cervical cancer cells through upregulation of T-cadherin. *Biomedicine & pharmacotherapy = Biomedecine & pharmacotherapie,* 107: 957-966.

[344] Wang, L. F., Mou, X. L., Xiao, L., and Tang, L. D. (2013). T-cadherin expression in uterine leiomyoma. *Arch Gynecol Obstet,* 288: 607-614.

[345] Chmelarova, M., Sirak, I., Mzik, M., Sieglova, K., Vosmikova, H., Dundr, P., Nemejcova, K., Michalek, J., Vosmik, M., Palicka, V., and Laco, J. (2016). Importance of Tumour Suppressor Gene Methylation in Sinonasal Carcinomas. *Folia biologica,* 62: 110-119.

[346] Sun, D., Zhang, Z., Van, D. N., Huang, G. W., Ernberg, I., and Hu, L. F. (2007). Aberrant methylation of CDH13 gene in

nasopharyngeal carcinoma could serve as a potential diagnostic biomarker. *Oral Oncol,* 43: 82-87.

[347] Misawa, K., Mochizuki, D., Imai, A., Endo, S., Mima, M., Misawa, Y., Kanazawa, T., Carey, T. E., and Mineta, H. (2016). Prognostic value of aberrant promoter hypermethylation of tumor-related genes in early-stage head and neck cancer. *Oncotarget,* 7: 26087-26098.

[348] Misawa, K., Kanazawa, T., Misawa, Y., Imai, A., Endo, S., Hakamada, K., and Mineta, H. (2011). Hypermethylation of collagen alpha 2 (I) gene (COL1A2) is an independent predictor of survival in head and neck cancer. *Cancer Biomark,* 10: 135-144.

[349] Wang, Q., Zhang, X., Song, X., and Zhang, L. (2018). Overexpression of T-cadherin inhibits the proliferation of oral squamous cell carcinoma through the PI3K/AKT/mTOR intracellular signalling pathway. *Archives of oral biology,* 96: 74-79.

[350] Takeuchi, T., Misaki, A., Liang, S. B., Tachibana, A., Hayashi, N., Sonobe, H., and Ohtsuki, Y. (2000). Expression of T-cadherin (CDH13, H-cadherin) in human brain and its characteristics as a negative growth regulator of epidermal growth factor in neuroblastoma cells. *J Neurochem,* 74: 1489-1497.

[351] Piperi, C., Themistocleous, M. S., Papavassiliou, G. A., Farmaki, E., Levidou, G., Korkolopoulou, P., Adamopoulos, C., and Papavassiliou, A. G. (2010). High incidence of MGMT and RARbeta promoter methylation in primary glioblastomas: association with histopathological characteristics, inflammatory mediators and clinical outcome. *Mol Med,* 16: 1-9.

[352] Lu, Y. T., Xiao, L. M., Liu, Y. W., Wang, H., Li, H., Zhou, Q., Pan, J., Lei, B. X., Huang, A., and Qi, S. T. (2015). MIR517C inhibits autophagy and the epithelial-to-mesenchymal (-like) transition phenotype in human glioblastoma through KPNA2-dependent disruption of TP53 nuclear translocation. *Autophagy,* 11: 2213-2232.

[353] Shi, Y., Wang, Y. Y., Luan, W. K., Wang, P., Tao, T., Zhang, J. X., Qian, J., Liu, N., and You, Y. P. (2014). Long Non-Coding RNA H19 Promotes Glioma Cell Invasion by Deriving miR-675. *PloS one,* 9.

[354] Suzuki, T., Wada, S., Eguchi, H., Adachi, J., Mishima, K., Matsutani, M., Nishikawa, R., and Nishiyama, M. (2013). Cadherin 13 overexpression as an important factor related to the absence of tumor fluorescence in 5-aminolevulinic acid-guided resection of glioma Laboratory investigation. *J Neurosurg,* 119: 1331-1339.

[355] Gutmann, D. H., Wu, Y. L., Hedrick, N. M., Zhu, Y., Guha, A., and Parada, L. F. (2001). Heterozygosity for the neurofibromatosis 1 (NF1) tumor suppressor results in abnormalities in cell attachment, spreading and motility in astrocytes. *Human Molecular Genetics,* 10: 3009-3016.

[356] Huang, Z. Y., Wu, Y., Hedrick, N., and Gutmann, D. H. (2003). T-cadherin-mediated cell growth regulation involves G2 phase arrest and requires p21(CIP1/WAF1) expression. *Mol Cell Biol,* 23: 566-578.

[357] Zhong, Y., Lopez-Barcons, L., Haigentz, M., Ling, Y. H., and Perez-Soler, R. (2004). Exogenous expression of H-cadherin in CHO cells regulates contact inhibition of cell growth by inducing p21 expression. *International journal of oncology,* 24: 1573-1579.

[358] Alkebsi, L., Handa, H., Yokohama, A., Saitoh, T., Tsukamoto, N., and Murakami, H. (2016). Chromosome 16q genes CDH1, CDH13 and ADAMTS18 are correlated and frequently methylated in human lymphoma. *Oncology letters,* 12: 3523-3530.

[359] Ogama, Y., Ouchida, M., Yoshino, T., Ito, S., Takimoto, H., Shiote, Y., Ishimaru, F., Harada, M., Tanimoto, M., and Shimizu, K. (2004). Prevalent hyper-methylation of the CDH13 gene promoter in malignant B cell lymphomas. *International journal of oncology,* 25: 685-691.

[360] Cosialls, A. M., Santidrian, A. F., Coll-Mulet, L., Iglesias-Serret, D., Gonzalez-Girones, D. M., Perez-Perarnau, A., Rubio-Patino, C., Gonzalez-Barca, E., Alonso, E., Pons, G., and Gil, J. (2012). Epigenetic profile in chronic lymphocytic leukemia using methylation-specific multiplex ligation-dependent probe amplification. *Epigenomics-Uk,* 4: 491-501.

[361] Uhm, K. O., Lee, E. S., Lee, Y. M., Park, J. S., Kim, S. J., Kim, B. S., Kim, H. S., and Park, S. H. (2009). Differential Methylation Pattern of ID4, SFRP1, and SHP1 between Acute Myeloid Leukemia and Chronic Myeloid Leukemia. *J Korean Med Sci,* 24: 493-497.

[362] Mu, H. J., Xie, P., Shen, Y. F., Jiang, Y. Q., and Zeng, Y. J. (2009). Cadherin-13 in primary and blast crisis chronic myeloid leukaemia: declining expression and negative correlation with the BCR/ABL fusion gene. *Brit J Biomed Sci,* 66: 20-24.

[363] Takeuchi, T., Misaki, A., Sonobe, H., Liang, S. B., and Ohtsuki, Y. (2000). Is T-cadherin (CDH13, H-cadherin) expression related to lung metastasis of osteosarcoma? *Histopathology,* 37: 193-194.

[364] Zucchini, C., Bianchini, M., Valvassori, B., Perdichizzi, S., Benini, S., Manara, M. C., Solmi, R., Strippoli, P., Picci, P., Carinci, P., and Scotlandi, K. (2004). Identification of candidate genes involved in the reversal of malignant phenotype of osteosarcoma cells transfected with the liver/bone/kidney alkaline phosphatase gene. *Bone,* 34: 672-679.

[365] Bromhead, C., Miller, J. H., and McDonald, F. J. (2006). Regulation of T-cadherin by hormones, glucocorticoid and EGF. *Gene,* 374: 58-67.

[366] Ma, H. S., Wang, E. L., Xu, W. F., Yamada, S., Yoshimoto, K., Qian, Z. R., Shi, L., Liu, L. L., and Li, X. H. (2018). Overexpression of DNA (Cytosine-5)-Methyltransferase 1 (DNMT1) And DNA (Cytosine-5)-Methyltransferase 3A (DNMT3A) Is Associated with Aggressive Behavior and Hypermethylation of Tumor Suppressor Genes in Human Pituitary Adenomas. *Med Sci Monitor,* 24: 4841-4850.

[367] Qian, Z. R., Sano, T., Yoshimoto, K., Asa, S. L., Yamada, S., Mizusawa, N., and Kudo, E. (2007). Tumor-specific downregulation and methylation of the CDH13 (H-cadherin) and CDH1 (E-cadherin) genes correlate with aggressiveness of human pituitary adenomas. *Modern Pathol,* 20: 1269-1277.

[368] Gustmann, S., Klein-Hitpass, L., Stephan, H., Weber, S., Bornfeld, N., Kaulisch, M., Lohmann, D. R., and Dunker, N. (2011). Loss at Chromosome Arm 16q in Retinoblastoma: Confirmation of the Association with Diffuse Vitreous Seeding and Refinement of the Recurrently Deleted Region. *Gene Chromosome Canc,* 50: 327-337.

[369] Gratias, S., Rieder, H., Ullmann, R., Klein-Hitpass, L., Schneider, S., Boloni, R., Kappler, M., and Lohmann, D. R. (2007). Allelic loss in a minimal region on chromosome 16q24 is associated with vitreous seeding of retinoblastoma. *Cancer research,* 67: 408-416.

[370] Marchong, M. N., Chen, D., Corson, T. W., Lee, C., Harmandayan, M., Bowles, E., Chen, N., and Gallie, B. L. (2004). Minimal 16q genomic loss implicates Cadherin-11 in retinoblastoma. *Mol Cancer Res,* 2: 495-503.

In: Cadherins
Editor: Jonathan McWilliam

ISBN: 978-1-53618-077-0
© 2020 Nova Science Publishers, Inc.

Chapter 2

CADHERIN-MEDIATED CELL ADHESION WITHIN THE SEMINIFEROUS TUBULES

Rita Payan-Carreira[1,*] *and Dario Santos*[2]
[1]MED - Mediterranean Institute for Agriculture,
Environment and Development & Dept. of Veterinary Medicine, ECT,
Universidade de Évora [Pole at Mitra], Évora, Portugal
[2]CITAB & Dept. of Biology and Environment,
Universidade de Trás-os-Montes e Alto Douro (UTAD),
Vila Real, Portugal

ABSTRACT

Cadherins (Cadh) are key-molecules in Adherens junctions (AJs). They are multiprotein complexes mediating cell-cell adhesion, and particularly important to shape cell polarity, provide plasticity and maintain architectural integrity. Cadh, a large superfamily of cell surface glycoproteins, present a unique extracellular region domain folding like the immunoglobulin domains. They are found in a wide array of species and a multitude of tissues, including the testis. In the mammalian testis, the seminiferous tubules represent a unique type of epithelium-like tissue,

---

* Corresponding Author's E-mail: rtpayan@gmail.com.

composed of two different cellular populations: the Sertoli somatic cells and the spermatogenic cells. Different sorts of cell-to-cell attachments connect adjacent Sertoli cells and Sertoli to germ cells. The overall arrangement of junctions forms the blood-testis barrier. These connections offer an immune-privileged environment to the developing germ cells, and the nutritional and metabolic support to germ cells while offering particular plasticity to the tubular structure. They allow the migration of differentiated germ cells from the basal towards the adluminal compartment while providing a tight-fitting barrier for paracellular translocation of molecules and particles. Between adjacent Sertoli cells, various types of homotypic adherens junctions exists, while heterotypic junctions are present between Sertoli and spermatogonia (basolateral junctions) or spermatid heads (apical junctions). Intercellular N-cadherin connections, anchored in cytoplasmic plaques involving (but not limited to) actin filaments, form different morphological types of AJs. All sorts of AJs work together with tight and gap junctions to form the blood-testis barrier. The integrity of the different adherens junctions are critical for the spermatogenic process and the production of viable spermatozoa. In this chapter, we propose to review and discuss the structure of the cadherin-mediated junctions in the seminiferous tubules and their function in male fertility.

Keywords: N-cadherin, Sertoli cells, germ cells, cell-cell connections, seminiferous tubules, spermatogenesis, fertility

ABBREVIATIONS

ADIP	afadin-and alpha-actinin-binding protein
aES	apical ectoplasmic specialization
AJs	adherens junctions
arp3	actin-related protein 3
N-WASP	neural Wiskott-Aldrich syndrome protein
AXPC	axial protocadherin
bES	basal ectoplasmic specialization
BTB	blood-testis barrier
Cadh	cadherins
CLMP	CXADR-like membrane protein precursor
CRB3	protein crumbs homolog 3 precursor

Ds	dachsous protein
Dsc	desmocollin
Dsg	desmoglein
Dvl	dishevelled
EC	extracellular cadherin
EGF	endotheliun growth factor
ERK1/2	extracellular signal-regulated kinase 1/2
ES	ectoplasmic specialization
FAK	focal adhesion/adherens kinase
Fmi	flamingo protein
Fz	Frizzled protein (receptor)
GJs	gap junctions
ILK	Integrin-linked kinase
JAM	junctional adhesion molecule
LMO7	LIM domain only protein 7
MAG	myelin associated glycoprotein
MAPK	mitogen-activated protein kinase
Pcdh	protocadherins
SPD	spermatids
SPG	spermatogonium
Src	proto-oncogene tyrosine-protein kinase
TBC	tubulobulbar complexes
TGF-β	transforming growth factor beta
TJs	tight junctions
TNF	tumour necrosis factor
WASP	Wiskott-Aldrich syndrome protein
Yes	tyrosine protein kinase
ZO	zonula occludens (protein)

INTRODUCTION

Epithelial cells found in many organs are characterized by their arrangement into cohesive sheets [1]. In tissue, strong cell-cell linkage is achieved by the called cell junctions and mediated by transmembrane proteins that are anchored intracellularly to the cytoskeleton [2, 3]. In vertebrates, cell junctions are classified into three major categories based on their physiological role: anchoring junctions, occluding junctions and communicating junctions [4]. Anchoring junctions are multiprotein complexes. They are critical components in the development and homeostasis of higher eukaryotic organisms allowing the adhesion of cells to each other and to large insoluble proteins of the extracellular matrix [5]. They are essential to shape cell polarity, provide plasticity and maintain architectural integrity helping the tissue to survive to inflicted mechanical stress [6]. At the anchoring junction complexes, cadherins play a crucial role as transmembrane adhesion proteins linking the cytoskeleton inside the cell to adjacent cell structures. When bonded to actin, the Cadh complex forms adherens junctions (AJs) while when bonded to intermediate filaments, it forms desmosomes [3]. The second category, named as occluding junctions, includes tight junctions (TJs) that create a physical barrier between adjacent epithelial cells controlling the movement and transport between the apical and basal layers of epithelia [7]. The last category of cell-cell junctions includes the gap junctions (GJs), a channel-like structure made up of proteins that allow the selective transport of material having molecular weight under 1000 Da between cells [4].

AJs are actin or intermediate filaments-based junctions. Cadherins (Cadh) and nectins are key-molecules in AJs. The expression of Cadh in the mammalian testis has been described in some reports. However, much controversy still exists regarding its localization and functions particularly on their involvement in the formation of the testis-specific adherent junctions. In this chapter, we intend to review the role of cadherin-mediated junctions in the seminiferous tubules and present a short description of AJs dynamics during spermatogenesis. The complexity of the structural junctions of the cells in the seminiferous tubules made it challenging to present a clear and concise description of the junction complexes existing there. The fact that those junctions interconnect, driving some conflicting reports on the literature, hardened this task. TJs and GJs overlap and integrate the adherence areas. The changes with time in the nomenclature of testicular junctions only add a defy to it. The information presented in this chapter does not exhaust the theme addressed herein but intends to serve as an opening for those starting on the topic.

JUNCTIONAL COMPLEXES IN THE SEMINIFEROUS TUBULES

In adult testicular cross-sections, the parenchyma shows a typical architecture: multiple segments of seminiferous tubules containing polarized Sertoli cells and the germinal epithelium, surrounded by a basement membrane and species-specific number of myoid cells layers; the tubules are separated by an interstitium of diverse density where Leydig cells embed [8].

In the mammalian testis, the seminiferous tubules represent a unique type of epithelium-like tissue [9], composed of two different cellular populations [10, 11]:

- The Sertoli somatic cells, in a basal position, are laterally anchored to each other. They provide nutrition and support to the developing spermatogenic cells, to which they are connected by multiple cell-to-cell attachment structures. These cells orchestrate the development and differentiation of germ cells by providing structural and metabolic support to germ cells and maintaining an immune-protective environment.
- The germ cell population, which can be found in different developmental stages, organized according to a regular cellular association (named stages of the seminiferous epithelium). Cells originating from a differentiated spermatogonium form a syncytium, maintained interconnected by intercellular bridges that derive from incomplete cytokinesis. This morphofunctional feature is of paramount importance to the synchronous development of germ cells and the success of spermatogenesis.

Sperm production occurs regularly in the post-pubertal testicular seminiferous tubules, according to a temporal species-specific pattern. To evolve normally, spermatogenesis demands an intense and complex interplay between germ cells and Sertoli cells as well as between the Sertoli and Leydig cells. Spermatogenesis aims to produce a genetically unique, haploid male cell from an originally diploid germ cell. For spermatogenesis to succeed, the developing germ cells depend on a tightly controlled environment, provided by the Sertoli cells [12]. In particular, seminiferous tubules are immune-privileged spaces that ensure the survival of advanced haploid germ cells (spermatids) that would otherwise be seen as foreign (non-self) by the immune system [13, 14]. The production and survival of haploid spermatozoa are paramount in male fertility.

Within the seminiferous tubules, Sertoli cells form an epithelial barrier, one of the tightest known blood-tissue barriers in the mammalian body. Named as the blood-testis barrier (BTB), it has been long recognized as crucial to protect the developing germ cells from the self-immune system [14]. BTB fosters the integrity of the seminiferous epithelium and the functional interconnections between the somatic and germ cells.

For long, it was considered the tight junctions between Sertoli cells composed BTB. Nonetheless, BTB is far more complex than initially thought. Today it is accepted that it does not rely exclusively on epithelial TJs, but rather represents a multifunctional system of junctions [15-17]. Therefore, the creation of an immune-privileged space in the seminiferous tubules is only one aspect of a larger role played by the junctional complexes in the seminiferous tubules. Whether or not the AJs should be included within the BTB junctions remained controversial. More recently, it has been demonstrated that, unlike other existing blood-tissue barriers, in the blood-testis barrier tight junctions (TJs) coexist and intermix with adherens junctions (AJs) and gap junctions (GJs) (Table 1), that work in concert to maintain BTB integrity and support spermatogenesis [14, 18-20].

The spermatogenic process is characterized by a continuously synchronized and spatially organized sequence of cell proliferation, differentiation, translocation and morphogenesis [21], whose success depends on the support and coordination of Sertoli cells. The spatial arrangement of germ cell within the tubules, resulting in the BTB particular junctional system, is crucial to spermatogenesis. Spermatogenesis depends on the intimate crosstalk between germ cells and Sertoli cells. It drives the nutritional and development support for each stage of the spermatogenic process, the timely transport of developing germ cells from near the basement membrane to the adluminal compartment of seminiferous tubules, as fully developed spermatids [12]. This arrangement demands a continuous remodelling of the junctional system to cope with the proliferation and differentiation of the germ cells and the transmigration of developing cells from one compartment to the other. The disassembly and reassembly of the local junctions when the germ cells move through the tubular compartments help to maintain the integrity of the barrier [18]. The different junctional complexes work together [16], as in a canal or barrage sluice lock to allow the passage of a boat through sections of different water levels. The plasticity of this system is crucial to male fertility, allowing the migration of differentiated germ cells from the basal towards the adluminal compartment, while at the same time providing a tight-fitting barrier for the paracellular translocation of molecules and particles (hence maintaining the strength of the blood-testis barrier) [10, 11].

As stated by Hess and Vogl [15], Sertoli cell anatomy embodies one of the most complex, three-dimensional structures in biology. Sertoli cells are polarized columnar cells that extend the entire height of the seminiferous epithelium, creating a barrier that protects and supports generations of developing non-somatic cells by creating a complex system of intricate membrane associations that is different between basal and apical locations [15].

Table 1. Classification of intercellular junctions within the seminiferous tubules [21-23]

Functional characterization	Occluding junctions	Anchoring junctions					Communicating junctions
		linked to actin filaments			linked to intermediated filaments		
Type of junctions	Tight junctions	Classic Adherens junctions	Ectoplasmic specializations	Tubulo-bulbar complexes	"Desmosome-like junctions"		Gap Junctions
Links Sertoli cells to — Adjacent Sertoli cells	√	√	√		√		√
Links Sertoli cells to — Germ cells		√ [stem SPG]			√		√
Links Sertoli cells to — Spermatids (SPD)			√	√ [elongated SPD]	√ [round SPD]		
Transmembrane proteins	Occludins; Claudins; JAM (A&B); CAR; CLMP; CRB3	Cadherins (N-E-P); Nectins (2-3) JAM (B-C)	Cadherins (N-E-P); Nectins (2-3); JAM-C; Vezatin	Clathrins	Desmogleins; Desmocollins; Connexins		Connexins
Adaptor proteins	ZO (2-3); MAGI; Cingulin	Catenin (α, β, γ..) Catenins Afadin	Catenins, Afadin, ZO-1	Dynamins (2-3); Cofilin; arp3; N-WASP	Plakophilins; Plakoglobins; Desmoplakins; Plectins		ZO-1
Main role	Compartmentalization of the testis	Tissue integrity					Communication

Within the different junctional complexes existing in the seminiferous tubules, AJs are the anchoring junctions connecting the cytoskeleton of adjacent cells, providing a robust tissue architecture. Testicular AJs link adjacent Sertoli cells, germ cells to Sertoli cells and germ cells to each other. This kind of structure provides mechanical support to the epithelium and allows germ cells movements through the different compartments of the tubules, besides playing a determinant role in germ cell morphogenesis and differentiation [24].

In the testis, classical cadherin-mediated AJs exists only between Sertoli cells and spermatogonia in the stem cells niche [16]. These junctions co-exist, however, with a modified type of adherens junctions called the ectoplasmic specialization (ES), and the tubulobulbar complexes (TBC), in a time- and cycle dependent pattern. ES established between adjacent Sertoli cells can be found in the basal compartment, and ES established between Sertoli cells and elongating spermatids were described within the adluminal compartment [21, 25, 26]. ES are unique, testis-specific AJs involving actin filaments, endoplasmic reticulum, and microtubules beneath the plasma membrane of Sertoli cells [19]. ES is expressed in the basal compartment (basal ES; bES), where it is intermixed with TJs, basal TBC and "desmosome-like" junctions at the BTB adjacent to the basement membrane, as well as in the adluminal compartment (apical ES), which engage both several TJ proteins (e.g., JAM-C) and focal adhesion complex (FAC) proteins [27]. Apical ES (aES) is the only junction type at the developing spermatid–Sertoli cell interface and seems to gather in one junctional type the ability to confer adhesion, communication, and cell movement and polarization [27]. TBC seems to be formed before spermiation, between Sertoli cells and the elongating spermatids. TBC are testis-specific actin-related endocytic structures that form between Sertoli cells and elongated spermatids at the point where the ectoplasmic specializations begin to disassemble [19]. All those types of junctions act in concert and show dynamic changes all through the gonad development as well as in a cyclic pattern during spermatogenesis [15, 28].

Domke [29], based on molecular studies, defends that due to their particular features, it is incorrect to use the name "desmosome-like junctions". Previous studies [9], showed that the AJs in seminiferous tubules contain N-cadherin connected to different adaptors (α- and β-catenin, plakoglobin, proteins p120, p0071 and a protein of the striatin family and, in rodents, also the proteins ZO-1 and myozap). According to the pattern and disposition of the junctions, the authors propose the use of the names "*areae adhaerentes*" for the larger N-Cadherin mediate junctions, and "*cribelliform junctions*" for the small clusters of strainer-like intercellular junctions perforated by cytoplasm-to-cytoplasm channels, localized close to the apical SE, found in bovine seminiferous tubules [9]. Notwithstanding, at present, no substantial evidence supports one or the other opposing opinion. Therefore, we retain the designation of "desmosome-like" junctions in this chapter.

CADHERINS

Cadherins are cell surface glycoproteins usually associated with cell-cell adhesion and recognition processes occurring in animal tissues. Cadh were discovery in the 1970s after unexpected observations of Masatoshi Takeichi in Japan [30]. Takeichi′ studies showed that cells interact using two adhesion systems, calcium-dependent and calcium-independent and that calcium protects the calcium-dependent adhesion molecule from proteolysis [31]. The name "cadherins" was introduced by Takeichi and stands for "calcium cell-cell adhesion molecules" [32]. Cadh forms a superfamily of more than one hundred of calcium-dependent membrane proteins [33]. Cadh are found both in vertebrates and invertebrates in a wide array of species and tissues [34, 35].

Cadh are intrinsic membrane proteins with an N-terminal extracellular region (ectodomain), followed by a single-pass α-helix transmembrane domain and a C-terminal intracellular region [36]. The extracellular cadherin (EC) domain consisting of about 110 residues folded into a sandwich of seven β-strands forming two β-sheets is the structural hallmark of Cadh family members [37, 38]. Cadh present from 2 to 34 ECs in a plethora of structural arrangements [38]. Ca^{2+} bounds to sites between adjacent EC domains inhibiting a free rotation around their linker peptides, thereby stabilizing cadherin domains together into rigid rods [39]. The number and homology of EC domains determine de division of Cadh superfamily of proteins into subfamilies: the classical (type I and II) Cadh, the desmosomal Cadh, the protocadherins (clustered and non-clustered Cadh), and the atypical Cadh (FAT, FAT-like or Dachsous and the large Cadh) [33, 40-42].

Classical Cadh do much more than hold cells together. They control cell movements underlying morphogenesis, changes in cell polarity, cell structure, and also mediate several intracellular signalling processes to control both cytoplasmic organization and motile behaviours of cells as well as changes in gene expression to control cell differentiation and growth and tissue architecture [41, 43]. Additionally, Cadh present physiological regulation properties in tissues, in a barrier and synaptic mechanisms [43]. The classical cadherins are subdivided into types I and II based on sequence homologies (e.g., E-, N-cadherins in type I, and VE-cadherin, in type II). They are involved in cell-to-cell adhesion and are concentrated at the adherens junctions. Classical (and desmosomal) Cadh ectodomain contains five EC domains usually designated as EC1 – EC5 beginning with the N-terminus of the molecule. These Cadh ectodomains usually form Ca^{2+}-dependent interactions with the Cadh ectodomains of adjacent cells (homophilic interactions) [36]. The cytoplasmic domain of classical Cadh links to catenins and, through catenins, to actin cytoskeleton filaments and intracellular signalling proteins [37, 43, 44].

Desmosomal cadherins [desmocollin (Dsc) or desmoglein (Dsg)] are typically found in desmosomes, intercellular junctions of epithelia and cardiac muscle. They are characterised by the enhanced stability and adhesive strength [45]. The types of desmosomal cadherins vary with the tissue, accounting for the tissue-specificity of desmosomes. Moreover, the role of these cadherins goes beyond those of cell-to-cell adhesion, and have been connected to intracellular signal transduction pathways controlling cell proliferation and differentiation [45].

Protocadherins (e.g., α-, β- and γ-Pcdhs; PAPC; AXPC) are the largest Cadh subfamily. Their structure is similar to type I classical cadherins but they possess an extracellular domain composed of six to seven EC repeats lacking conserved sequence elements present in classical cadherins. Additionally, the intracellular domain of protocadherins (Pcdh) is structurally diverse and do not interact with catenins, but instead with other proteins such as Fyn-kinase [46]. Only vertebrates expressed Pcdh, mostly in the nervous system, but can also be expressed elsewhere [34, 47]. Yet, their functions remain unclear. It has been hypothesized that non-clustered Pcdh promote cell motility and migration, mediate dendrite ramification (axon growth and patterning) and dendritic self-avoidance (self- and non-self-recognition of neurons). They also may regulate synapse dynamics [48], thereby participating in the neural circuit assembly.

The atypical Cadhs Dachsous (Ds), Fat and Flamingo (Fmi) control the cellular and tissue organisation by regulating growth and planar cell polarity signalling [40, 49]. Ds and Fat have 27 and 34 ECs, respectively [50]. Fmi is the only cadherin with a seven-pass, rather than a single, transmembrane domain, and also possess an extracellular sequence including nine (Drosophila) or eight (mouse and human) ECs [41, 49, 51]. This group of planar polarity proteins further includes the Wingless (or Wnt) armadillo (the prototype of the family that includes β-catenin and plakoglobin), Frizzled (Fz) and Dishevelled (Dvl). A recently identified human homologue (EGFL2) has nine cadherin repeats [40, 42].

CADHERIN IN THE JUNCTIONAL COMPLEXES OF THE SEMINIFEROUS TUBULES

AJs are actin or intermediate filaments-based junctions. Cadherins have classically been considered the primary adhesive unit at adherens junctions. By linking to underlying actin belts via catenins, Cadh confers resistance to the intercellular adhesion. Cadh also engages in numerous functions other than their participation in the adhesive complex in AJs, including cell-cell recognition, cytoskeletal organization, signal transduction, acquisition of cell polarity and growth control [41]. In multiple body structures, Cadh often cooperate with nectins in AJs. The association between the nectin and cadherin systems is physically mediated by afadin, α-catenin, and their binding proteins, namely ponsin, LMO7, ADIP, vinculin, or α-actinin [52].

In the seminiferous tubules, AJs are junctional complexes based on N-Cadherin anchored in a diversity of cytoplasmic plates. Seminiferous tubules AJs (whether ES, TBC or the formerly named "desmosome-like" junctions) are composed of integral-membrane proteins, adaptors, and signalling molecules. The integral membrane proteins in testicular AJs encompass cadherins, nectins, and integrins. These molecules use β-catenin, α-actinin, afadin, vinculin, or cofilin as adaptors to the cellular cytoskeleton of actin (ES and TBC) and other intermediate filaments (vimentin in the "desmosome-like junctions") or microtubules (TBC) [53].

Cadherins Presence in the Seminiferous Tubules

Different cadherins have been identified in the testicular tissue in some species. Classical Cadh have been identified within the seminiferous tubules of fetal, immature and adult testis, each exhibiting a unique expression profile during the testicular development [54-57]. The time- and spatially orchestrated changes of the gonad development as well as with the stage of the spermatogenesis in mature tubules suggest a crucial role to be played by these molecules in the testicular homeostasis. Accordingly, Cadh inhibition with specific antibodies in the adult gonad entrains the disassembly of the tubular architecture and leads to the progressive loss of germ cells [19]. It has also been proposed cadherins to play an essential role during collective cell migration [58], by acting as a mechanosensor connected with the contractile actomyosin cytoskeleton via α-catenin and vinculin [58, 59]. This role is mediated through conformational changes in α-catenin under tension that expose vinculin binding sites [59, 60]. Vinculin recruitment to cell–cell junctions increases the duration of the interaction between Cadh-catenin complex and the actin cytoskeleton, fostering AJs stabilization during intercellular tensive stress [60].

The expression of E-Cadh seems to be restricted to the fetal and immature testis before the formation of BTB [19, 55]. It has been proposed to be involved in the migration of primordial germ cells and the gonad differentiation, particularly in the formation of testicular cords and the survival of germ cells [61]. However, its presence after puberty has been elusive. It has been shown that E-Cadh is only expressed by spermatogonial stem cells of adult mice [62] and rats [63], and peripubertal sheep [64].

Early in the nineties, N-Cadh was found to be expressed by both the Sertoli and germ cells and recognized as one of the molecules participating in the cellular junctions at the seminiferous epithelium [65]. It is now acknowledged that N-Cadh is present in testes at all the developmental stages. Its expression and localization obey a tight spatiotemporal regulation [55, 57]. In the post-pubertal seminiferous tubules, N-Cadh (colocalized with β-catenin and p120) displays a stage-specific immunostaining pattern. It locates in the basal compartment of the seminiferous tubules, close but independently of the basal ectoplasmic specializations [19, 66], suggesting its participation in the "desmosome-like" junctions. It was also found in association with spermatogonia, spermatocytes and the heads of elongated spermatids [66, 67]. Albeit N-Cadh was also located around the round spermatids, evidence showed that the junctional complexes here lack plectin or β-catenin adaptor proteins, suggesting that a different junctional complex is formed after the dismantling and reassembly of the Sertoli-germ cell junction as spermatocytes develop into round spermatids [54] and migrates into the adluminal compartment. This finding supported the hypothesis that the N-Cadh/β-catenin complexes, anchored to the actin network, regulates cell adhesive function together with the "desmosome-like" junctions linking Sertoli and other germ cells [68].

Besides, N-Cadh is present at either the basal and apical ES. In the former, the anchoring complex is established with some proteins [69] usually seen in the focal adhesion complex at the cell-matrix interface (such as vinculin, FAK, claudins and ZO-1). Thereby, basal ES may be considered as a hybrid cell-cell anchoring junction [20]. Conversely, apical ES is also considered to be a hybrid intercellular actin-based AJs that integrates N-cadh/catenin complexes [70], yet it is composed of proteins customarily found within tight junctions and focal contacts, such as α6β1 integrin or the nectin/afadin complex [69]. Moreover, its relation with the typical actin filaments is limited to the Sertoli cell side [71].

The expression of Cadh other than the N-cadherin are controversial and no consensual information is retrieved from studies using immune-based techniques; also, more recent reports often contradict results from previous ones. This controversy might be explained by the differences in the antibodies used for detection. Additionally, the existence of species particularities or the existence of a short, transitory expression pattern may also account for the reported differences in respect to the expression of cadherins or its adaptors as well as the unavailability of appropriate antibodies against specific Cadh epitopes. Below we present a summary of the data provided or validated by gene expression studies.

Cdh-6 immunostaining was also described in the adult rat seminiferous tubules. It is invariably colocalized with p120 at Sertoli-round spermatid desmosome-like junctions [54], and the heads of elongated spermatids, in a transient stage-related pattern. Contrasting with N-Cadh, no catenins other than p120 were found associated with Cadh-6. Still, these findings were not confirmed in the studies by Domke [23].

P-cadherin has been localized in the testis, but only in the junctions between adjacent peritubular myoid cells [21], and therefore will not be addressed in this chapter.

More recently, VE-Cadh has been localized in the seminiferous tubules of adult mice, exclusively in germ cells, in contrast to what has been described so far for the other Cadhs in the seminiferous tubules [72]. VE-Cadh has been located in round spermatids and in early elongating spermatids, before the formation of apical ES, for which Berruti and colleagues [72] defend that its role may be related to the spermatid acquisition of polarity or the orientation inside the seminiferous tubule. As VE-Cadh is completely internalized in elongated spermatids, these authors further argue that it may not a component of a fully established aES. In the same study, Berruti and colleagues [72] suggested that VE-Cadh could be involved in the junctions linking Sertoli cells and round spermatids ("desmosome-like" junctions).

Cadherin Adaptors and Scaffold Proteins

In canonic AJs, β-catenin bridges Cadh to α-catenin, which in turn conjugates the cadh/catenin complex to the actin cytoskeleton or vimentin filaments in mature Sertoli cells [68]. α-catenin is crucial to the strength of AJs, and facilitates the linkage of the Cadh-catenin complex to the actin cytoskeleton [68] if the interaction is under tension [59].

Membrane-associated β-catenin and p120 catenin partnering with cell-specific cadherins (including E-Cadh and N-Cadh) were localized in the fetal gonads at the time of gonad differentiation [73]. In the adult testis, expression of Cadh-associated catenins (β-catenin, α-catenin and p120) have been associated with basal ES, at the junctions between Sertoli cells and spermatogonia [21], primary spermatocytes [19, 54], but also in tight junctions between adjacent Sertoli cells [16]. The N-Cadh-β-catenin complex at the basal ES is crucial for BTB maintenance [74]. In the "desmosome-like" junctions, Cadh-catenin complex also integrates γ-catenin (known also as plakoglobin) [9]. Even though the complex N-Cadh-β-catenin has been evidenced at the apical ES, the existence of α-catenin has not been demonstrated there, suggesting that β-catenin may interact with an alternative adaptor in those particular junctions [68].

Besides its participation in the cadherin/catenin complexes, α-catenin may also bind to afadin, vinculin and zyxin, to link with the actin cytoskeleton [27, 75]. Nectins anchors to other proteins, such as catenins or afadin, thereby clustering Cadhs and stabilizing the AJs [76]. In the testis, this arrangement between the nectin/afadin and the cadherin/catenin complex allows for the crosstalk between the two types of adhesion complexes [24], fostering a type of AJs that are at the same time strong and easily remodeled [76]. This interaction is particularly significant for the later stages of the development of spermatids and during the assembly and disassembling of various types of intercellular adhesion [21].

It has been shown that in the testis, the assembly of AJs and GJs are interdependent. The connection between the two types of junctions is provided by the fact that N-Cadh associated β-catenin colocalize with connexin 43, which favours a close physical and functional association between AJs and GJs [77]. Similarly, afadin and catenins mediate interactions with ZO proteins, thus providing another chain to the link between TJs and AJs [78].

There were also additional peripheral proteins regulating the actin dynamics, such as zyxin, axin and WASP (Wiskott-Aldrich syndrome protein), shown to be structurally linked with N-cadh/β-catenin complexes in the seminiferous tubules [19, 79], suggesting that N-cadh/β-catenin complexes may also indirectly link vimentin-based intermediate filaments via these adaptors. Additionally, c-Src protein and c-Yes have been identified as proteins peripheral to N-cadherin [27], and proposed as possible mediators of the regulatory system that modulates the dynamics of the Sertoli–germ cell AJs.

The studies by Domke [23] failed to evidence desmocollin or desmoglein in the seminiferous tubules of different species.

Alternative Adaptors and Scaffolds for N-Cadherin in the Seminiferous Tubules

In the seminiferous tubules, adaptors other than the catenins have been related to apical ES (aES), a hybrid anchoring junction type sharing the properties of the different junctional complexes found in the basal compartment of the seminiferous tubules [27]. In aES, N-cadh has been associated to α6-β1 integrin, linking to γ3-laminin [21, 80] that together with laminin α3 and β3 forms a functional integrin binding protein in elongating spermatids [27]. The α6-β1 integrin is expressed in Sertoli cells, while α3/β3/γ3-laminins are expressed in the elongating spermatids [81]. In aES integrin-laminin complexes connecting to the actin cytoskeleton, the strength of the junction is lower compared to the bES, coping with the changes in the shape of elongating spermatids and facilitating an easy disengagement of the cells at spermiation.

This integrin-laminin functional complex is supported by the co-existence of other peripheral regulatory proteins, such as the integrin-linked kinase (ILK), focal adhesion kinase (FAK), p-FAK, Src, vinculin, and paxillin [27].

Interactions between the cadherin- and the integrin-mediated AJs may occur through different adaptors and scaffold (e.g., p120, α- or γ-catenin, vinculin) or effector proteins (e.g., non-receptor tyrosine kinases such as Src and Yes, or Focal Adherens Kinases – FAK, phosphatases) [82]. Integrin and Cadh mediated junctions share several common signaling components, e.g., vinculin or FAT [83]. Also, data from other tissues showed that Necls (nectin-like molecules) interact with nectins and associates with integrins and Growth Factor Receptor [78]. α3-β1 integrin associates with Cadh-mediated AJs in multiple tissues. Tyrosine phosphorylation of β-catenin, mediated by α3-β1 integrin via the tetraspanin interaction, shifts the molecule from the Cadh complexes. Freed, β-catenin becomes available to act in other pathways such as the Wnt. Also, through the phosphorylation of α-catenin, modulates the Cadh complex connection to the actin cytoskeleton [84]. Integrins and Cadh complexes are also intrinsically linked to each other via their binding to the actin-myosin network [83].

JUNCTION DYNAMICS AND SPERMATOGENESIS

Through spermatogenesis, developing germ cells cross the BTB and pass from the basal into the adluminal tubular compartment to finally be released as fully maturated spermatids into the lumen of the tubules (spermiation). The main events requesting the turnover of AJs are the transport of pre-leptotene/leptotene spermatocyte across the BTB, the movement of round spermatids back and forth the seminiferous epithelium, and spermiation [16, 85].

During this process, the junctional complexes mediating cell to cell interaction in the seminiferous tubules (either between Sertoli cells and between the Sertoli and germ cells) suffer changes and are remodelled, on a precise spatiotemporal dependent regulation [86]. The transcompartmental migration implies the disassembly and reassembly not only of the TJs but also of the AJs up- and downstream the migrating cells. To the perfect synchrony of the junctional remodelling, it is of utmost importance the existence of functional crosstalk as well as a fine-tuned mechanism of recycling within and between compartments, a phenomenon intimately dependent on GJs [16]. Similarly, before spermiation, the apical ES suffers rearrangements, a transitory AJs complex forms – the TBC -, in a process tightly coordinated. The TBC has been associated with a "recycling device" allowing the Sertoli cell to reuse structures and molecules belonging to former/changed junctional complexes. Signals originating in one type of junctions will determine the changes in the other junctional complexes elsewhere in the cell, which is achieved by changes in membrane trafficking, cytoskeleton association or binding affinity [82].

The assembling-disassembling of AJs is a dynamic and stringently regulated process, responding to intra- or extracellular stimuli [75]. The regulation of the AJs may be achieved through transcriptional and post-transcriptional mechanisms directed to the adhesion complex as well as to the actin cytoskeleton. The transcriptional mechanism regulates the *de novo* production of any element of the junctional complexes. In a structure like the seminiferous tubules, where an immense number of cells develop and need to migrate across and out of the tubules, the homeostasis of the process should not depend exclusively on de synthesis of new molecules to rebuild the junctions, the posttranslational mechanisms allowing the recycling and trafficking of molecules to assemble new AJs predominate [16]. A cell can control the protein activity and its interactions at a given place via posttranslational modifications (e.g., glycosylation, ubiquitination, proteolysis, and phosphorylation) [87].

AJs dynamics in the seminiferous tubules are mainly regulated by protein kinases and phosphatases that phosphorylate the AJ structural proteins [88], thereby breaking down the complex N-Cadh-β-catenin and consequently its connection to the actin cytoskeleton. Changes in the linkage to the actin cytoskeleton would also assist germ cell propulsion into a new location.

It has been proposed that the SRC family of non-receptors protein kinases are essential players in the remodelling of the junctional complexes during intratubular cell migration, representing a critical hub controlling junctional remodelling. Among those non-receptor proteins, c-Src and c-Yes are present in either Sertoli and germ cells. They trigger the phosphorylation of adhesion cell proteins [85, 89]. In the seminiferous tubule, c-Src structurally associates with p-FAK (Focal Adhesion Kinase) and the integrin-laminin complex. On the other hand, c-Yes interacts with occludin, FAK, N-cadherin, β- and γ-catenin, β1-integrin as well as with actin (via the Epidermal Growth Factor receptor-8) [90]. An increase in tyrosine residues of proteins phosphorylation at the cadherin/catenin complexes drives a loss of interaction with the actin cytoskeleton, and thereby to the loss of intercellular adhesion. However, these molecules regulate far more than the dynamics of AJs, the availability of the junctional proteins or the actin interaction. It has been shown that they also play a role in the endocytic vesicle-mediated protein trafficking [91] in Sertoli cells, that allows the recycling of molecules of the junctional complexes during the restructuring of AJs. Endosome vesicles with internalized cadh or their adaptor proteins might be transported across the Sertoli cell to where the new junctions are assembled [85], even if in a different tubular compartment, or be sent to degradation. c-Src and c-Yes seems to have opposing roles in the outcome of disassembled proteins, the c-Yes driving the molecules towards the endocytic vesicles trafficking mechanism (transcytosis/recycling), whereas c-Src fosters their lysosomal degradation [85, 91].

On the other hand, the reversible ubiquitination of adhesion molecules will also foster degradation of the molecules (in proteasomes), namely occludins, Cadhs and β-catenin, disassembling TJs and AJs. Ubiquitinating and deubiquitinating enzymes have been identified in the seminiferous tubules, and propose to contribute to changes in the permeability of junctional complexes [92].

Several molecular regulators have been implicated in the AJs dynamics, namely cytokines (e.g., TGF-β or TNF) and hormones (e.g., testosterone). Androgens are important determinants of spermatogenesis. It has been shown that testosterone plays a crucial role during the restructuring of the junctional complex in the seminiferous tubules [93], by mediating the mechanisms of endocytosis, transcytosis and recycling of the membrane-linked proteins that participate in the intercellular interface of tubular junctions [16]. In particular, testosterone actions on the AJs dynamics are exerted thought nonclassical signalling cascades, via c-Src and MAP-Kinase [86, 89]. Testosterone-induced phosphorylation of EGF receptor II, activates SRC family proteins [89], which promote the recycling of integrated membrane proteins. Contrastingly, cytokines enhance endosome-mediated degradation. Moreover, ERK1/2 activation by the TGF-β3 receptor induces AJs disruption, in a MAPK-mediated pathway that recruits peripheral adaptors. Those molecules (e.g., ZO-1 and α- and γ-catenin) allow the linkage of the TJs- and AJs-integral membrane proteins, which do not physically interact with each other, thereby contributing to the coordinated junctional remodelling in the seminiferous tubules [92, 94].

An additional candidate participating in the AJs remodelling is the Nitric Oxide (NO)-cGMP signalling pathway [95], which regulates β-catenin availability to partner with N-Cadherin [96], driving the dissolution of the AJ structural protein complexes.

CONCLUSION

In the seminiferous tubules, Cadherins are important actors in the interplay of the AJs, TJs and GJs that together form the blood-testis-barrier. BTB is one of the most complex form of cell-to-cell connection, and a fundamental structure requesting a finely tuned control to support the development and protection of the spermatogenic epithelium. This brief review intent to highlight the role of N-Cadh in testicular AJs and its role in the junctional dynamics. Limited literature is available at the topic, and most references address the role of AJs as a complement for the TJs in BTB. Much remains elusive regarding the mechanisms regulating AJs during spermatogenesis, and on how they behave during the germ cells movement across the epithelium. A few research groups are now working with hypothetical models which may prove useful to design future functional experimental studies whether to support the quest for male fertility and contraceptive purposes. The ever-growing pace of the science undoubtedly will soon provide new insights on Cadhs and AJs in the seminiferous tubules.

REFERENCES

[1] Honda, H. 2017. "The world of epithelial sheets." *Dev Growth Differ* 59 (5):306-316. doi: 10.1111/dgd.12350.

[2] Obrink, B. 1986. "Epithelial cell adhesion molecules." *Exp Cell Res* 163 (1):1-21. doi: 10.1016/0014-4827(86)90554-9.

[3] Wei, Q., and H. Huang. 2013. "Insights into the role of cell-cell junctions in physiology and disease." *Int Rev Cell Mol Biol* 306:187-221. doi: 10.1016/B978-0-12-407694-5.00005-5.

[4] Alberts, B., A. Johnson, J. Lewis, D. Morgan, M. Raff, K. Roberts, and P. Walter. 2015. "Cell junctions, Cell adhesion and the extracelular matrix." In *Molecular Biology of the Cell*, edited by G. S. Lewis and E. Zayatz, 1035-1090. New York: Garland Science.

[5] LaFlamme, S. E., and P. A. Vincent. 2006. "Cell Junctions, Structure, Function, and Regulation." *Reviews in Cell Biology and Molecular Medicine*. doi: 10.1002/3527600906.mcb.200300165.

[6] Niessen, C. M. 2007. "Tight junctions/adherens junctions: basic structure and function." *J Invest Dermatol* 127 (11):2525-32. doi: 10.1038/sj.jid.5700865.

[7] Pollard, T. D., W. C. Earnshaw, J. Lippincott-Schwartz, and G. T. Johnson. 2017. "Chapter 30 - Cellular Adhesion." In *Cell Biology (Third Edition)*, edited by Thomas D. Pollard, William C. Earnshaw, Jennifer Lippincott-Schwartz and Graham T. Johnson, 525-541. Elsevier.

[8] Ni, F. D., S. L. Hao, and W. X. Yang. 2019. "Multiple signaling pathways in Sertoli cells: recent findings in spermatogenesis." *Cell Death Dis* 10 (8):541. doi: 10.1038/s41419-019-1782-z.

[9] Domke, L. M., S. Rickelt, Y. Dörflinger, C. Kuhn, S. Winter-Simanowski, R. Zimbelmann, R. Rosin-Arbesfeld, H. Heid, and W. W. Franke. 2014. "The cell-cell junctions of mammalian testes: I. The adhering junctions of the seminiferous epithelium represent special differentiation structures." *Cell Tissue Res* 357 (3):645-65. doi: 10.1007/s00441-014-1906-9.

[10] Lara, N. L. M., G. M. J. Costa, G. F. Avelar, S. M. S. N. Lacerda, R. A. Hess, and L. R. de França. 2018. "Testis Physiology—Overview and Histology." In *Encyclopedia of Reproduction (Second Edition)*, edited by Michael K. Skinner, 105-116. Oxford: Academic Press.

[11] Lara, N. L. M., G. F. de Avelar, G. M. J. Costa, S. M. S. N. Lacerda, R. A. Hess, and L. R. de França. 2018. "Cell–Cell Interactions—Structural." In *Encyclopedia of Reproduction (Second Edition)*, edited by Michael K. Skinner, 68-75. Oxford: Academic Press.

[12] Goldberg, E., and B. R. Zirkin. 2018. "Spermatogenesis: Overview." In *Encyclopedia of Reproduction (Second Edition)*, edited by Michael K. Skinner, 13-18. Oxford: Academic Press.

[13] Stanton, P. G. 2016. "Regulation of the blood-testis barrier." *Semin Cell Dev Biol* 59:166-173. doi: 10.1016/j.semcdb.2016.06.018.

[14] Mao, B., M. Yan, L. Li, and C. Y. Cheng. 2018. "Blood-Testis Barrier." In *Encyclopedia of Reproduction (Second Edition)*, edited by Michael K. Skinner, 152-160. Oxford: Academic Press.

[15] Hess, R. A., and A. W. Vogl. 2015. "Sertoli cell anatomy and cytoskeleton." In *Sertoli Cell Biology (Second Edition)*, edited by Michael D. Griswold, 1-55. Oxford: Academic Press.

[16] Yan Cheng, C., and D. D. Mruk. 2015. "Biochemistry of Sertoli cell/germ cell junctions, germ cell transport, and spermiation in the seminiferous epithelium." In *Sertoli Cell Biology (Second Edition)*, edited by Michael D. Griswold, 333-383. Oxford: Academic Press.

[17] Cheng, C. Y., and D. D. Mruk. 2009. "Regulation of blood-testis barrier dynamics by focal adhesion kinase (FAK): an unexpected turn of events." *Cell Cycle* 8 (21):3493-9. doi: 10.4161/cc.8.21.9833.

[18] Mruk, D. D., and C. Y. Cheng. 2015. "The Mammalian Blood-Testis Barrier: Its Biology and Regulation." *Endocr Rev* 36 (5):564-91. doi: 10.1210/er.2014-1101.

[19] Vazquez-Levin, M. H., C. I. Marín-Briggiler, J. N. Caballero, and M. F. Veiga. 2015. "Epithelial and neural cadherin expression in the mammalian reproductive tract and gametes and their participation in fertilization-related events." *Dev Biol* 401 (1):2-16. doi: 10.1016/j.ydbio.2014.12.029.

[20] Yan, H. H. N., D. D. Mruk, W. M. Lee, and C. Y. Cheng. 2008. "Cross-Talk between Tight and Anchoring Junctions—Lesson from the Testis." In *Molecular Mechanisms in Spermatogenesis,* edited by C. Yan Cheng, 234-254. New York, NY: Springer New York.

[21] Goossens, S., and F. van Roy. 2005. "Cadherin-mediated cell-cell adhesion in the testis." *Front Biosci* 10:398-419. doi: 10.2741/1537.

[22] Ravel, C., and S. Jaillard. 2011. "[The Sertoli cell]." *Morphologie* 95 (311):151-8. doi: 10.1016/j.morpho.2011.07.118.

[23] Domke, L. M. 2018. *Molecular and ultrastructural characteristics of adhering junctions and cytoskeletons in cells of mammalian testes.* Doctor of Natural Sciences Dissertation, Combined Faculty of Natural Sciences and Mathematics, Ruperto Carola University, Germany.

[24] Lui, W.-Y., D. D. Mruk, W. M. Lee, and C. Y. Cheng. 2003. "Adherens Junction Dynamics in the Testis and Spermatogenesis." *Journal of Andrology* 24 (1):1-14. doi: 10.1002/j.1939-4640.2003.tb02627.x.

[25] Berruti, G., and C. Paiardi. 2014. "The dynamic of the apical ectoplasmic specialization between spermatids and Sertoli cells: the

case of the small GTPase Rap1." *Biomed Res Int* 2014:635979. doi: 10.1155/2014/635979.

[26] Toyama, Y., M. Maekawa, and S. Yuasa. 2003. "Ectoplasmic specializations in the Sertoli cell: new vistas based on genetic defects and testicular toxicology." *Anat Sci Int* 78 (1):1-16. doi: 10.1046/j.0022-7722.2003.00034.x.

[27] Wong, E. W., D. D. Mruk, and C. Y. Cheng. 2008. "Biology and regulation of ectoplasmic specialization, an atypical adherens junction type, in the testis." *Biochim Biophys Acta* 1778 (3):692-708. doi: 10.1016/j.bbamem.2007.11.006.

[28] França, L. R., R. A. Hess, J. M. Dufour, M. C. Hofmann, and M. D. Griswold. 2016. "The Sertoli cell: one hundred fifty years of beauty and plasticity." *Andrology* 4 (2):189-212. doi: 10.1111/andr.12165.

[29] Domke, L. M. 2019. "The cell-cell junctions of mammalian testes-a summary." *Cell Tissue Res*. doi: 10.1007/s00441-019-03150-3.

[30] Takeichi, M. 2018. "Historical review of the discovery of cadherin, in memory of Tokindo Okada." *Dev Growth Differ* 60 (1):3-13. doi: 10.1111/dgd.12416.

[31] Takeichi, M. 1977. "Functional correlation between cell adhesive properties and some cell surface proteins." *J Cell Biol* 75 (2 Pt 1):464-74. doi: 10.1083/jcb.75.2.464.

[32] Yoshida-Noro, C., N. Suzuki, and M. Takeichi. 1984. "Molecular nature of the calcium-dependent cell-cell adhesion system in mouse teratocarcinoma and embryonic cells studied with a monoclonal antibody." *Dev Biol* 101 (1):19-27. doi: 10.1016/0012-1606(84)90112-x.

[33] Colás-Algora, N., and J. Millán. 2019. "How many cadherins do human endothelial cells express?" *Cell Mol Life Sci* 76 (7):1299-1317. doi: 10.1007/s00018-018-2991-9.

[34] Nollet, F., P. Kools, and F. van Roy. 2000. "Phylogenetic analysis of the cadherin superfamily allows identification of six major subfamilies besides several solitary members." *J Mol Biol* 299 (3):551-72. doi: 10.1006/jmbi.2000.3777.

[35] Oda, H., and M. Takeichi. 2011. "Evolution: structural and functional diversity of cadherin at the adherens junction." *J Cell Biol* 193 (7):1137-46. doi: 10.1083/jcb.201008173.

[36] Pokutta, S., and W. I. Weis. 2007. "Structure and mechanism of cadherins and catenins in cell-cell contacts." *Annu Rev Cell Dev Biol* 23:237-61. doi: 10.1146/annurev.cellbio.22.010305.104241.

[37] Angst, B. D., C. Marcozzi, and A. I. Magee. 2001. "The cadherin superfamily: diversity in form and function." *J Cell Sci* 114 (Pt 4):629-41.

[38] Hulpiau, P., I. S. Gul, and F. van Roy. 2016. "Evolution of Cadherins and Associated Catenins." In *The Cadherin Superfamily: Key Regulators of Animal Development and Physiology*, edited by Shintaro T. Suzuki and Shinji Hirano, 13-37. Tokyo: Springer Japan.

[39] Shapiro, L., and W. I. Weis. 2009. "Structure and biochemistry of cadherins and catenins." *Cold Spring Harb Perspect Biol* 1 (3):a003053. doi: 10.1101/cshperspect.a003053.

[40] Wheelock, M. J., and K. R. Johnson. 2003. "Cadherins as modulators of cellular phenotype." *Annu Rev Cell Dev Biol* 19:207-35. doi: 10.1146/annurev.cellbio.19.011102.111135.

[41] Halbleib, J. M., and W. J. Nelson. 2006. "Cadherins in development: cell adhesion, sorting, and tissue morphogenesis." *Genes Dev* 20 (23):3199-214. doi: 10.1101/gad.1486806.

[42] Brasch, J., O. J. Harrison, B. Honig, and L. Shapiro. 2012. "Thinking outside the cell: how cadherins drive adhesion." *Trends Cell Biol* 22 (6):299-310. doi: 10.1016/j.tcb.2012.03.004.

[43] Gumbiner, B. M. 2016. "Classical Cadherins." In *The Cadherin Superfamily: Key Regulators of Animal Development and Physiology*, edited by Shintaro T. Suzuki and Shinji Hirano, 41-69. Tokyo: Springer Japan.

[44] Heuberger, J., and W. Birchmeier. 2010. "Interplay of cadherin-mediated cell adhesion and canonical Wnt signaling." *Cold Spring Harb Perspect Biol* 2 (2):a002915. doi: 10.1101/cshperspect.a002915.

[45] Chidgey, M., and D. Garrod. 2016. "Desmosomal Cadherins." In *The Cadherin Superfamily: Key Regulators of Animal Development and Physiology*, edited by Shintaro T. Suzuki and Shinji Hirano, 159-193. Tokyo: Springer Japan.

[46] Junghans, D., I. G. Haas, and R. Kemler. 2005. "Mammalian cadherins and protocadherins: about cell death, synapses and processing." *Curr Opin Cell Biol* 17 (5):446-52. doi: 10.1016/j.ceb.2005.08.008.

[47] Sano, K., H. Tanihara, R. L. Heimark, S. Obata, M. Davidson, T. St John, S. Taketani, and S. Suzuki. 1993. "Protocadherins: a large family of cadherin-related molecules in central nervous system." *EMBO J* 12 (6):2249-56.

[48] Hayashi, S., and M. Takeichi. 2015. "Emerging roles of protocadherins: from self-avoidance to enhancement of motility." *J Cell Sci* 128 (8):1455-64. doi: 10.1242/jcs.166306.

[49] Fulford, A. D., and H. McNeill. 2019. "Fat/Dachsous family cadherins in cell and tissue organisation." *Curr Opin Cell Biol* 62:96-103. doi: 10.1016/j.ceb.2019.10.006.

[50] Sharma, P., and H. McNeill. 2013. "Fat and Dachsous cadherins." *Prog Mol Biol Transl Sci* 116:215-35. doi: 10.1016/B978-0-12-394311-8.00010-8.

[51] Usui, T., Y. Shima, Y. Shimada, S. Hirano, R. W. Burgess, T. L. Schwarz, M. Takeichi, and T. Uemura. 1999. "Flamingo, a seven-pass transmembrane cadherin, regulates planar cell polarity under the control of Frizzled." *Cell* 98 (5):585-595.

[52] Fujiwara, T., A. Mizoguchi, and Y. Takai. 2016. "Cooperative Roles of Nectins with Cadherins in Physiological and Pathological Processes." In *The Cadherin Superfamily: Key Regulators of Animal Development and Physiology*, edited by Shintaro T. Suzuki and Shinji Hirano, 115-156. Tokyo: Springer Japan.

[53] Lee, N. P., and C. Y. Cheng. 2004a. "Adaptors, junction dynamics, and spermatogenesis." *Biol Reprod* 71 (2):392-404. doi: 10.1095/biolreprod.104.027268.

[54] Johnson, K. J., and K. Boekelheide. 2002. "Dynamic testicular adhesion junctions are immunologically unique. II. Localization of classic cadherins in rat testis." *Biol Reprod* 66 (4):992-1000. doi: 10.1095/biolreprod66.4.992.

[55] Munro, S. B., and O. W. Blaschuk. 1996. "A comprehensive survey of the cadherins expressed in the testes of fetal, immature, and adult mice utilizing the polymerase chain reaction." *Biol Reprod* 55 (4):822-7. doi: 10.1095/biolreprod55.4.822.

[56] Byers, S. W., S. Sujarit, B. Jegou, S. Butz, H. Hoschutzky, K. Herrenknecht, C. MacCalman, and O. W. Blaschuk. 1994. "Cadherins and cadherin-associated molecules in the developing and maturing rat testis." *Endocrinology* 134 (2):630-9. doi: 10.1210/endo.134.2.7507830.

[57] Rode, K., H. Sieme, P. Richterich, and R. Brehm. 2015. "Characterization of the equine blood-testis barrier during tubular development in normal and cryptorchid stallions." *Theriogenology* 84 (5):763-72. doi: 10.1016/j.theriogenology.2015.05.009.

[58] Ebnet, K., D. Kummer, T. Steinbacher, A. Singh, M. Nakayama, and M. Matis. 2018. "Regulation of cell polarity by cell adhesion receptors." *Semin Cell Dev Biol* 81:2-12. doi: 10.1016/j.semcdb.2017.07.032.

[59] Padmanabhan, A., M. V. Rao, Y. Wu, and R. Zaidel-Bar. 2015. "Jack of all trades: functional modularity in the adherens junction." *Curr Opin Cell Biol* 36:32-40. doi: 10.1016/j.ceb.2015.06.008.

[60] Collins, C., and W. J. Nelson. 2015. "Running with neighbors: coordinating cell migration and cell-cell adhesion." *Curr Opin Cell Biol* 36:62-70. doi: 10.1016/j.ceb.2015.07.004.

[61] Piprek, R. P., M. Kolasa, D. Podkowa, M. Kloc, and J. Z. Kubiak. 2019. "Tissue-specific knockout of E-cadherin (Cdh1) in developing mouse gonads causes germ cells loss." *Reproduction.* doi: 10.1530/REP-18-0621.

[62] Tolkunova, E. N., A. B. Malashicheva, E. V. Chikhirzhina, E. I. Kostyleva, W. Zeng, J. Luo, I. Dobrinski, A. Hierholzer, R. Kemler, and A. N. Tomilin. 2009. "E-cadherin as a novel surface marker of

spermatogonial stem cells." *Cell and Tissue Biology* 3 (2):103-109. doi: 10.1134/S1990519X09020011.

[63] Zhang, Y., H. Su, F. Luo, S. Wu, L. Liu, T. Liu, B. Yu, and Y. Wu. 2011. "E-cadherin can be expressed by a small population of rat undifferentiated spermatogonia in vivo and in vitro." *In Vitro Cellular & Developmental Biology - Animal* 47 (8):593. doi: 10.1007/s11626-011-9446-z.

[64] Zhang, Y., S. Wu, F-h. Luo, Baiyinbatu, L-h. Liu, T-y. Hu, B. Yu, G-p. Li, and Y-j. Wu. 2014. "CDH1, a Novel Surface Marker of Spermatogonial Stem Cells in Sheep Testis." *Journal of Integrative Agriculture* 13 (8):1759-1765. doi: 10.1016/S2095-3119(13)60689-9.

[65] Newton, S. C., O. W. Blaschuk, and C. F. Millette. 1993. "N-cadherin mediates Sertoli cell-spermatogenic cell adhesion." *Dev Dyn* 197 (1):1-13. doi: 10.1002/aja.1001970102.

[66] Johnson, K. J., and K. Boekelheide. 2002. "Dynamic testicular adhesion junctions are immunologically unique. I. Localization of p120 catenin in rat testis." *Biol Reprod* 66 (4):983-91. doi: 10.1095/biolreprod66.4.983.

[67] Andersson, A. M., K. Edvardsen, and N. E. Skakkebaek. 1994. "Expression and localization of N- and E-cadherin in the human testis and epididymis." *Int J Androl* 17 (4):174-80. doi: 10.1111/j.1365-2605.1994.tb01239.x.

[68] Lee, N. P., D. Mruk, W. M. Lee, and C. Y. Cheng. 2003. "Is the cadherin/catenin complex a functional unit of cell-cell actin-based adherens junctions in the rat testis?" *Biol Reprod* 68 (2):489-508. doi: 10.1095/biolreprod.102.005793.

[69] Mruk, D. D., and C. Y. Cheng. 2004. "Sertoli-Sertoli and Sertoli-germ cell interactions and their significance in germ cell movement in the seminiferous epithelium during spermatogenesis." *Endocr Rev* 25 (5):747-806. doi: 10.1210/er.2003-0022.

[70] Lie, P. P., C. Y. Cheng, and D. D. Mruk. 2011. "The biology of the desmosome-like junction a versatile anchoring junction and signal transducer in the seminiferous epithelium." *Int Rev Cell Mol Biol* 286:223-69. doi: 10.1016/B978-0-12-385859-7.00005-7.

[71] Mok, K-W., P. P. Y. Lie, D. D. Mruk, J. Mannu, P. P. Mathur, B. Silvestrini, and C. Y. Cheng. 2013. "The Apical Ectoplasmic Specialization-Blood-Testis Barrier Functional Axis is A Novel Target for Male Contraception." In *Biology and Regulation of Blood-Tissue Barriers,* edited by C. Yan Cheng, 334-355. New York, NY: Springer New York.

[72] Berruti, G., M. Ceriani, and E. Martegani. 2018. "Dynamic of VE-cadherin-mediated spermatid-Sertoli cell contacts in the mouse seminiferous epithelium." *Histochem Cell Biol* 150 (2):173-185. doi: 10.1007/s00418-018-1682-9.

[73] Fleming, A., N. Ghahramani, M. X. Zhu, E. C. Délot, and E. Vilain. 2012. "Membrane β-catenin and adherens junctions in early gonadal patterning." *Developmental Dynamics* 241 (11):1782-1798. doi: 10.1002/dvdy.23870.

[74] Islam, R., H. Yoon, B. S. Kim, H. S. Bae, H. R. Shin, W. J. Kim, W. J. Yoon, Y. S. Lee, K. M. Woo, J. H. Baek, and H. M. Ryoo. 2017. "Blood-testis barrier integrity depends on Pin1 expression in Sertoli cells." *Sci Rep* 7 (1):6977. doi: 10.1038/s41598-017-07229-1.

[75] D'Souza-Schorey, C. 2005. "Disassembling adherens junctions: breaking up is hard to do." *Trends Cell Biol* 15 (1):19-26. doi: 10.1016/j.tcb.2004.11.002.

[76] Miyoshi, J., and Y. Takai. 2007. "Nectin and nectin-like molecules: biology and pathology." *Am J Nephrol* 27 (6):590-604. doi: 10.1159/000108103.

[77] Li, M. W. M., D. D. Mruk, and C. Y. Cheng. 2013. "Gap Junctions and Blood-Tissue Barriers." In *Biology and Regulation of Blood-Tissue Barriers*, edited by C. Yan Cheng, 260-280. New York, NY: Springer New York.

[78] Togashi, H. 2016. "Differential and Cooperative Cell Adhesion Regulates Cellular Pattern in Sensory Epithelia." *Front Cell Dev Biol* 4:104. doi: 10.3389/fcell.2016.00104.

[79] Lee, N. P., D. D. Mruk, A. M. Conway, and C. Y. Cheng. 2004. "Zyxin, axin, and Wiskott-Aldrich syndrome protein are adaptors that link the cadherin/catenin protein complex to the cytoskeleton at adherens junctions in the seminiferous epithelium of the rat testis." *J Androl* 25 (2):200-15. doi: 10.1002/j.1939-4640.2004.tb02780.x.

[80] Lee, N. P., and C. Y. Cheng. 2004b. "Ectoplasmic specialization, a testis-specific cell-cell actin-based adherens junction type: is this a potential target for male contraceptive development?" *Hum Reprod Update* 10 (4):349-69. doi: 10.1093/humupd/dmh026.

[81] Wu, S., M. Yan, R. Ge, and C. Y. Cheng. 2019. "Crosstalk between Sertoli and Germ Cells in Male Fertility." *Trends Mol Med.* doi: 10.1016/j.molmed.2019.09.006.

[82] Weber, G. F., M. A. Bjerke, and D. W. DeSimone. 2011. "Integrins and cadherins join forces to form adhesive networks." *J Cell Sci* 124 (Pt 8):1183-93. doi: 10.1242/jcs.064618.

[83] Mui, K. L., C. S. Chen, and R. K. Assoian. 2016. "The mechanical regulation of integrin-cadherin crosstalk organizes cells, signaling and forces." *J Cell Sci* 129 (6):1093-100. doi: 10.1242/jcs.183699.

[84] Chattopadhyay, N., Z. Wang, L. K. Ashman, S. M. Brady-Kalnay, and J. A. Kreidberg. 2003. "alpha3beta1 integrin-CD151, a component of the cadherin-catenin complex, regulates PTPmu expression and cell-cell adhesion." *J Cell Biol* 163 (6):1351-62. doi: 10.1083/jcb.200306067.

[85] Xiao, X., Y. Yang, B. Mao, C. Y. Cheng, and Y. Ni. 2019. "Emerging Role for SRC family kinases in junction dynamics during spermatogenesis." *Reproduction.* doi: 10.1530/REP-18-0440.

[86] Lui, W.-Y., and C. Y. Cheng. 2013. "Transcriptional Regulation of Cell Adhesion at the Blood-Testis Barrier and Spermatogenesis in the Testis." In *Biology and Regulation of Blood-Tissue Barriers*, edited by C. Yan Cheng, 281-294. New York, NY: Springer New York.

[87] Bertocchi, C., M. Vaman Rao, and R. Zaidel-Bar. 2012. "Regulation of adherens junction dynamics by phosphorylation switches." *J Signal Transduct* 2012:125295. doi: 10.1155/2012/125295.

[88] Chen, Y. M., N. P. Lee, D. D. Mruk, W. M. Lee, and C. Y. Cheng. 2003. "Fer kinase/FerT and adherens junction dynamics in the testis: an in vitro and in vivo study." *Biol Reprod* 69 (2):656-72. doi: 10.1095/biolreprod.103.016881.

[89] Chojnacka, K., and D. D. Mruk. 2015. "The Src non-receptor tyrosine kinase paradigm: New insights into mammalian Sertoli cell biology." *Mol Cell Endocrinol* 415:133-42. doi: 10.1016/j.mce.2015.08.012.

[90] Xiao, X., D. D. Mruk, F. L. Cheng, and C. Y. Cheng. 2013. "c-Src and c-Yes are Two Unlikely Partners of Spermatogenesis and their Roles in Blood-Testis Barrier Dynamics." In *Biology and Regulation of Blood-Tissue Barriers,* edited by C. Yan Cheng, 295-317. New York, NY: Springer New York.

[91] Xiao, X., Y. Ni, C. Yu, L. Li, B. Mao, Y. Yang, D. Zheng, B. Silvestrini, and C. Y. Cheng. 2018. "Src family kinases (SFKs) and cell polarity in the testis." *Semin Cell Dev Biol* 81:46-53. doi: 10.1016/j.semcdb.2017.11.024.

[92] Lui, W. Y., and C. Y. Cheng. 2007. "Regulation of cell junction dynamics by cytokines in the testis: a molecular and biochemical perspective." *Cytokine Growth Factor Rev* 18 (3-4):299-311. doi:10.1016/j.cytogfr.2007.04.009.

[93] Meng, J., R. W. Holdcraft, J. E. Shima, M. D. Griswold, and R. E. Braun. 2005. "Androgens regulate the permeability of the blood-testis barrier." *Proc Natl Acad Sci U S A* 102 (46):16696-700. doi: 10.1073/pnas.0506084102.

[94] Yan, H. H., and C. Y. Cheng. 2005. "Blood-testis barrier dynamics are regulated by an engagement/disengagement mechanism between tight and adherens junctions via peripheral adaptors." *Proc Natl Acad Sci U S A* 102 (33):11722-7. doi: 10.1073/pnas.0503855102.

[95] Mruk, D. D., O. Sarkar, and P. P. Mathur. 2008. "Nitric Oxide-cGMP Signaling: Its Role in Cell Junction Dynamics During Spermatogenesis." *Current Medicinal Chemistry - Immunology, Endocrine & Metabolic Agents)* 8 (1):28-35. doi: 10.2174/187152208783790741.

[96] Lee, N. P., D. D. Mruk, C. H. Wong, and C. Y. Cheng. 2005. "Regulation of Sertoli-germ cell adherens junction dynamics in the testis via the nitric oxide synthase (NOS)/cGMP/protein kinase G (PRKG)/beta-catenin (CATNB) signaling pathway: an in vitro and in vivo study." *Biol Reprod* 73 (3):458-71. doi: 10.1095/biolreprod.105.040766.

BIOGRAPHICAL SKETCH

Dario Joaquim Simões Loureiro dos Santos

Present Position: Assistant Professor

Affiliation: Department of Biology and Environment, University of Trás-os-Montes e Alto Douro (UTAD), Vila Real, Portugal

Education:

- Biology (licenciate) - University of Coimbra, Coimbra - Portugal
- Cellular and Molecular Biology (PhD) University of Coimbra, Coimbra - Portugal

Research and Professional Experience: Cellular Biology and Biochemistry, Cell Physiology, Phytochemicals, Oxidative Stress and antioxidants.

Rita Payan CArreira

Present Position: Full Professor

Affiliations: Mediterranean Institute for Agriculture, Environment and Development & Dept. of Veterinary Medicine, ECT, Universidade de Évora [Pole at Mitra], Évora, Portugal

Education:

- DVM - High School of Veterinary Medicine, Lisbon Technical University- Portugal
- Veterinary Sciences / Animal Reproduction (PhD) University of Trás-os-Montes e Alto Douro (UTAD), Vila Real, Portugal

Research and Professional Experience: Veterinary Theriogenology, Embryo maternal-crosstalk, Fertility determinants, Testicular pathology in small animals, Reproductive Physiology, Molecular biology.

In: Cadherins
Editor: Jonathan McWilliam
ISBN: 978-1-53618-077-0
© 2020 Nova Science Publishers, Inc.

Chapter 3

DISRUPTION OF E-CADHERIN PATTERN IN UTERINE AND MAMMARY TUMOURS

*Adelina Gama[1,2], Fernanda Seixas[1,2],
Maria dos Anjos Pires[1,2,]*, Fernando Schmitt[3,4]
and Rita Payan-Carreira[5]*

[1]CECAV, Universidade de Trás-os-Montes e Alto Douro (UTAD),
Vila Real, Portugal
[2]ECAV, Dept. Veterinary Sciences, Universidade de
Trás-os-Montes e Alto Douro (UTAD), Vila Real, Portugal
[3]IPATIMUP, Institute of Molecular Pathology and Immunology of the
University of Porto, Porto, Portugal
[4]Faculty of Medicine of the University of Porto, Porto, Portugal
[5]MED - Mediterranean Institute for Agriculture, Environment and
Development & Dept. of Veterinary Medicine, ECT,
Universidade de Évora [Pole at Mitra], Évora, Portugal

* Corresponding Author's E-mail: apires@utad.pt

Abstract

E-cadherin (E-cadh), a member of the classic cadherins superfamily, plays an important role in epithelial cell-to-cell adhesion, encompassing the dynamic interactions between adjacent cells including the control of morphogenesis, maintenance of cell polarity and tissue architecture. Cadherins comprise a large family of cell surface glycoproteins, presenting unique extracellular regions domains known as the cadherin motifs or domains, which fold like immunoglobulin domains. They mediate strong Ca^{2+}-dependent homophilic interactions between neighbouring epithelial cells, resulting in the formation of cell adhesion "zippers." E-cadh cytoplasmic tail links to catenins and, thereby to the actin cytoskeleton and signalling proteins to form a cell-cell signalling centre: it regulates several intracellular signal transduction pathways, including Wnt/β-catenin, PI3K/Akt, Rho GTPase, and NF-KB signalling.

E-cadh plays a crucial role in the barrier formation of polarized epithelial cell layers at the interfaces contacting with the external environment, namely the uterus and the mammary gland. The maintenance of these barriers could be considered as a prime immunologic function of E-cadh, compartmentalizing potentially harmful agents away from the underlying tissue. Disruption of classical cadherin expression has been related to the occurrence of diseases driving disturbances in tissue architecture, such as inflammation and cancer.

In cancer, loss of E-cadh expression/function increases cell proliferation, cell migration, and disruption of epithelial cell homeostasis, driving cell dissociation and scattering. Alterations of E-cadh expression have also been reported during particular moments of the female reproductive physiology, namely throughout the oestrous cycle or during the embryo-maternal interaction at embryo implantation (early pregnancy). Several studies have shown the downregulation of E-cadh in malignant epithelial tumours, which has been associated with loss of cell differentiation, epithelial to mesenchymal transition (EMT) and invasion. Data also suggest that loss of E-cadh may be associated with malignant progression, metastasis, and reduced survival in multiple cancer patients.

In this chapter, we review and discuss the role of E-cadh in the uterine and mammary gland homeostasis and describe disruptive patterns of E-cadh expression in neoplastic conditions of the uterus and mammary glands in human and domestic dogs and cats.

Keywords: E-cadherin, uterus, mammary gland, carcinoma, human, dog, cat

INTRODUCTION

Cadherins are cell surface glycoproteins generally associated with cell-to-cell adhesion and recognition processes occurring in animal tissues. Cadherins form a superfamily of more than one hundred of calcium-dependent membrane proteins mediating homophilic cell-to-cell adhesion [1]. Presenting unique extracellular regions domains, known as the cadherin motifs or cadherin domains, which fold like immunoglobulin domains [2], these proteins have been found in both invertebrates and vertebrates, and in a wide array of tissues [3, 4].

Yet, classic cadherin roles go far beyond the epithelial cell-to-cell adhesion. They control cell movements underlying morphogenesis, changes in cell polarity, cell structure, and also mediate several intracellular signalling processes associated with cytoplasmic organization and motile behaviours of cells, as well as changes in gene expression to control cell differentiation and growth, and tissue architecture [5, 6]. Cadherins contribute to tissue homeostasis, participating in the tissue barrier function, cell proliferation, and migration [7]. As other type I classic cadherins, E-cadherin (E-cadh) mediates cell adhesion and is a key determinant of epithelial morphology and differentiation in most epithelial body tissues and organs [8].

E-Cadherin at the Core of Epithelial Adherens Junctions

Epithelial integrity depends on the interaction of different types of junctions, namely the tight junctions, adherens junctions and desmosomes. Together, they constitute the epithelial junctional complex. Adherens junctions (AJs) are specialized cell-to-cell adhesion sites consisting of E-cadh/catenin complexes that link to actin cytoskeleton [6], which in turn regulates the assembling, organization, stability and remodeling of AJs [9]. Connection with actin bundles allows the interaction of other types of junctions with AJs, working together to maintain the epithelial barrier. In the case of AJs, the barrier is achieved by creating a continuous adhesive belt at the apical–lateral interface of adjacent cells. Despite providing strength and polarity to epithelial barriers, AJs present notable plasticity, which is important for tissue morphology and morphogenesis [10]. This plasticity is of upmost importance to coordinated multicellular movement or the cooperatively regulated single-cell migration [11, 12], either for embryo morphogenesis, tissue healing and tumour invasion/progression.

As said, clusters of dimeric E-cadh mediate trans-homophilic interactions between neighboring cells forming AJs [10, 13]. The plasticity of the epithelial barrier is made possible by a rapid and constant turnover of membrane E-cadh, driving its internalization through different endocytic pathways and its recycling by exocytosis [13]; in contrast, the *de novo* synthesis of E-cadh is relatively slow. The E-cadh internalization process is mediated by clathrin, which allows the fenomenon to be spacialy controled [10].

Inside the epithelial cells, the E-cadh cytoplasmic domain links to β-catenin and p120-catenin; via β-catenin, E-cadh links to α-catenin and thereby to actin [9, 13, 14]. α-catenin links to actin through either the F-actin or vinculin [13]. The binding of α-catenin to F-actin provides stability to the adhesion point, while p120-catenin stabilises the cell-to-cell adhesion by controlling the retention of E-cadh at the cell surface [9]. While linked to p120-catenin, E-cadh is guarded from clathrin-mediated endocytosis, driving the notion that p120-catenin acts as a master regulator of cadherin stability [10, 12]. The fixation of E-cadh in the cell membrane results from the inhibition of endocytosis [13]. Besides its control of cadherin internalization, p120-catenin is considered as a "set-point" for cadherin expression [15]. It is now accepted that the level of E-cadherin expression, not that of the catenins, is the rate-limiting step for the formation of E-cadh complexes and the cell adhesion [16]. It has been proposed that p120-catenin regulates the cadherin levels within the cell, and possibly the switching of the pattern of cadherin expression from one member to another, specially in cells expressing multiple cadherin types. Moreover, it has been shown that p120-catenin may shuttle to the nucleus, where it interacts with the transcription factor Kaiso, which is involved with the regulation of various cancer-related genes [17].

Depletion of extracellular calcium drives the disruption of adhesion between adjacent epithelial cells [10]. Remodelling of AJs implies the control of E-cadh internalization, and thereby the disturbance of the E-cadh/catenin complexes. The disruption of these complexes is regulated mainly by protein kinases and phosphatases that phosphorylate the AJs structural proteins [13], affecting the connection of E-cadh to the actin cytoskeleton. Small GTPases, such as the RhoGTPases, Rac and Cdc42, and corresponding effectors, have been associated to AJs remodelling [9, 10, 13]. Furthermore, it has also been proposed that Src family of non-receptors protein kinases (e.g., c-Src and c-Yes) are essential players in AJs remodelling [18]. The activation of Src kinases drives the phosphorylation of AJ componentes and the subsequente disruption of cell-to-cell adhesions, therefore contributing to cell migration.

E-Cadherin as a Key Modulator during Development and Neoplastic Progression

Cadherin endocytosis plays a significant role in physiological and pathological processes. Cadherin endocytosis is crucial in the embryo differentiation and development [5], as well as in canine trophoblast migration at implantation [19]. A similar event occurs during epithelial-mesenchymal transition (EMT), where loss of membrane E-cadh (either by cadherin endocytosis or cadherin switching) favour proliferation, migration and invasion of neoplastic cells [15, 20].

Cadherin internalization may occur through different endocytic pathways, the clathrins' being the most studied [10, 15]. Additional pathways associated to cadherin internalization include the caveolin-mediated and macropinocytosis-like pathways. An increase in the phosphorylation of cadherin/catenin complexes drives a loss of interaction with the actin cytoskeleton, the internalization of membrane cadherin, and the loss of intercellular adhesion [21]. Nevertheless, not all the molecules entering an endocytic pathway undergo lysosomal degradation. Some cadherin molecules are recycled back to the cell membrane, allowing cells to retain their barrier properties and their polarity. The fate of E-cadh trafficking within the cell (degradation *vs.* recycling) is influenced by a family of Src non-receptors protein kinases [16]. c-Src and c-Yes seem to have opposing roles in the outcome of the complex disassembled proteins: c-Yes drives the molecules towards the endocytic vesicles trafficking mechanism (recycling back to the membrane), whereas c-Src fosters their lysosomal degradation [16, 22]. In some tumours, Src induced down-regulation of E-cadherin depends on integrin signaling and FAK phosphorylation [23]. Increased neoplastic Src activity have been associated with tumour invasiveness [16] and the loss of membrane E-cadh expression which suppresses cell-to-cell contacts and favours cell migration, invasiveness and metastatic dissemination. Therefore, E-cadh is often used as a prognostic marker for several solid tumours [16, 24].

Dysregulation of the E-cadh complexes may drive cells into the transition to a different cell phenotype, contributing to EMT which is often associated with tumour progression and metastases. During EMT, epithelial cells lose polarity and the typical intercellular adhesion, and develop mesenchymal features together with increased migratory and invasive properties [20]. However, EMT does not occur only in cancer; it can also be observed in natural biological phenomena such as during embryogenesis, wound healing [25], or even during glandular cyclic branching (e.g, the branching of the mammary ducts) [26].

Although initial studies pointed to a reductionist binary process, EMT is now proposed as a highly coordinated plastic program, manifesting as dynamic transitional states between the epithelial and mesenchymal phenotypes [27, 28], with the identification of distinct intermediate states (epithelial, intermediate hybrid and mesenchymal states) [27, 29].

EMT is not a mandatory pre-requisite for tumour cell invasion and metastasis [23, 30]. In fact, neoplastic cells can move in a collective manner, on which depend on the disturbance of cadherin-mediated cell-to-cell contacts. This invasive behaviour is also determined by p120-catenin, and may be accomplished by either an E-cadh-dependent or independent mechanism. Down-regulation of E-cadh can be achieved by the regulation of Rac and Cdc42 activity or indirectly by inhibiting RhoGTPases and downstream cytoskeletal dynamics involved in cell invasion and migration [31].

Cadherin switching, that alters intercellular adherence and the cadherin-associated signaling pathways, is a common event during EMT. While normal epithelial cells express particular patterns of cadherins, cells undergoing EMT start expressing different cadherin types (mesenchymal cadherins), such as Neuronal (N-) cadh, Placental (P-) Cadh, or cadherin 11, among others [20]. This switch seldom occurs at once, but follows a specific pattern, possibly responding to different internal or external stimuli. TGF-β and the Snail, Slug and Twist transcription factors have been proposed as main EMT regulators. Dysregulation of EMT may ultimately drive to cancer [25]. Cadherin switch may coordinate changes in several cell functions, namely in cell metabolism, resistance to hypoxia and programmed cell death mechanisms, and lower adhesion to extracellular matrices [32], which all contribute to proliferation and the invasive behaviour of neoplastic cells.

The loss of functional E-cadh will contribute to tumour cells resistance to apoptosis mediated by either the intrinsic or the extrinsic pathways [33-36]. More recently, Capra and Eskelinen [37] showed that the disruption of E-cadh mediated adhesion triggers the up-regulation of survivin, which suppresses apoptosis and promotes cell proliferation and migration; survivin signalling may therefore play an important role in neoplastic progression in cells not suffering EMT.

E-CADHERIN EXPRESSION IN UTERINE CANCER

E-Cadherin in the Normal Endometrium

The cadherin/catenin complex is an essencial intercellular adhesive system in the mammal endometrium, where it coordinates key morphogenetic processes, regulates epithelial differentiation and proliferation, and supports the epithelial phenotype [38]. Moreover, it also mediates the interactions between the embryo and maternal tissues at implantation [19, 39, 40].

The few available studies on the expression of E-cadh and/or β-catenin have established the existence of cyclic variations of these molecules throughout the uterine cycle and early pregnancy in humans [41-43], sheep [44], pigs [45] and dogs [19, 46].

Albeit studies on gene expression and protein location or quantification are not always concordant, it has been proposed that progesterone determines a decline in the strength of AJs which would facilitate the trophoblast invasion across the epithelial barrier [47]. This hypothesis is supported by a decrease in E-cadh around the implantation time in ewes [44], sows [48], and dogs [19]. In the feline cyclic endometrium, a similar decrease in the intensity of E-cadh membrane labelling has been found during progesterone dominance (M.A. Pires, unpublished results). Dudley et al. [49] described a dislocalization of E-cadh from the lateral plasma membrane to the cytoplasm in uterine epithelial cells in early pregnant cats.

Furthermore, studying the canine embryo-maternal interactions at implantation Payan-Carreira et al. [19] showed that internalization of E-cadh occurs in the maternal superficial epithelial layers of the endometrium, while the invading trophoblast retain the membrane labelling. This work agrees with an involvement of E-cadh/catenin complex in the embryo implantation and further suggests that the adhesiveness strength favours the collective invasion of trophoblastic cells.

E-Cadherin in Endometrial Carcinomas

The loss of cellular polarity [50] and alterations in cell adhesion [38] are hallmarks characteristics of cancer. Multiple studies reported reduced or aberrant expression of E-cadh and/or catenins in different human epithelial cancers: thyroid and esophageal carcinoma [38], breast carcinoma [38, 51, 52], gastric and pancreatic carcinoma, bladder and prostatic carcinoma, among others [38]. Defects in the E-cadh/catenin adhesion complex have been described in endometriosis [43] and in several gynecologic carcinomas, including ovarian, endometrial [38, 52, 53] and cervical carcinomas [38]. Endometrial cancer is the most common malignancy of the women genital tract. In the majority of cases the prognosis is good, but women with poor differentiated, deep myoinvasive tumours, or with extension of disease to other organs or lymph nodes within the pelvis, have frequently disease recurrence [54].

In endometrial cancer, E-cadh is a recognized putative marker of good prognosis [52, 55]. Loss of E-cadh expression, has been shown to drive the loss of cell adhesiveness, and contact inhibition [52], promoting tumour progression and an aggressive behaviour, invasion and metastasis in several epithelial tumours [52, 56]. In endometrial cancer, it is correlated with tumour dedifferentiation [50, 57], high grade histology [50, 54, 56], and deep myometrial invasion [50, 54, 57], higher rate of extrapelvic recurrence [54], presentation of other adverse prognostic factors and lower overall survival [52, 56]. Some studies showed that E-cadh expression is an independent prognostic factor for women endometrial carcinoma [54, 55]. Decreased membranous E-cadh expression is predictive for endometrial cancer mortality, disease progression, and extrapelvic recurrence, independent of known prognostic factors such as stage, grade, and histological subtype [54].

Feline endometrial adenocarcinomas are uncommon, poorly characterized lesions [58], that may be underdiagnosed [59], affecting even young cats [60, 61]. There are few published studies on the immunophenotype of these lesions [58] and with small case series [59-61] which makes the full characterization of these neoplasms difficult. Though, some studies reported E-cadh and β-catenin expression in both the normal feline endometrium [58], and in endometrial adenocarcinoma [58, 62]. According to Gil da Costa et al. [58], the loss of cell adhesion that occurs within these tumours does not require down-regulation of E-cadh expression; in addition, the nuclear translocation of β-catenin was not a characteristic feature of feline endometrial carcinomas.

In their study, Carico et al. [38] showed that human endometrial carcinomas may present different patterns of E-cadh and α-catenin expression within the tumour, reflecting the intratumoral heterogeneity of the neoplastic epithelium. Similarly, feline endometrial carcinomas present a patchy pattern of the cadherin/catenin complexes within the tumour, presenting areas of membrane and areas of cytoplasmic E-cadh expression (M.A. Pires, unpublished results). It is possible that while remaining anchored to their neighbours due to E-cadh membranar expression, epithelial neoplastic cells might have an impaired ability to metastasize. This hypothesis deserves to be further explored in the case of feline endometrial carcinomas, where a benign clinical course has been frequently described [60, 63, 64].

Canine endometrial carcinomas are rare, and mostly occur in old bitches [65]. Despite both E-cadh and β-catenin expression have been reported in the normal cyclic endometrium and the early pregnant uterus [19], the available literature is sparce in studies of adhesion molecule expression on canine endometrial carcinomas. This gap on information on E-cadh and other adhesion molecules on promotion, progression and prognosis of feline and canine endometrial carcinomas shows the need of new studies on this topic to evaluate the real effect of adhesion molecules on uterus tumourigenesis.

E-CADHERIN EXPRESSION IN MAMMARY CANCER

E-Cadherin Expression in Normal Mammary Tissues

As stated above, E-cadh has been described in most epithelial tissues, including the mammary gland. In normal human breast and in canine and feline mammary gland, E-cadh is expressed by luminal epithelial cells at the cell membrane [66-69]; this pattern of expression is also observed for α- and β-catenins [70-73]. In contrast, P-cadh is restricted to the basal myoepithelial cells [74-76]. Although cadherins expression in the mammary differentiated cells is well defined and cell-type specific, their expression in progenitor or mammary stem cell populations remain unclear [26]. Basal, P-cadh expressing progenitor cells seems to be responsible for mammary formation and branching morphogenesis, while alveolar progenitor cells most likely represent luminal cells which give rise to E-cadh positive cells [26, 77]. It has been proven that E-cadh is essential for mammary gland, after conditional E-cadh inactivation studies using knockout mouse models [26, 78].

E-Cadherin Expression in Mammary Cancer

Breast cancer is among the deadliest malignancies in developed countries [79], with the metastatic spread being the primary reason of this fatal outcome [79-81]. In canine species, spontaneous mammary tumours are the second most frequent tumour, and the most common neoplasia in the female dog. Regarding its biological behaviour, malignant cases account for up to 50% of female mammary tumours [82, 83]. In feline species, mammary neoplasias are among the most commonly diagnosed tumours in female cats [82, 84-86], accounting for 12% of all tumours and 17% of the tumours in queens [82]. Most studies reported that more than 80% of feline mammary tumours are malignant, along with rapid progression and metastasis [84, 86].

E-cadh, a putative tumour suppressor gene implicated in carcinogenesis [87, 88], is classically considered a good prognostic marker in cancer [89]; several studies proposed E-cadh as a tumour and invasion suppressor molecule, as invasion and metastasis are promoted when its expression is lost [70, 80, 90].

In human breast cancer, although representing one of the cancer types for which E-cadh has been extensively investigated for diagnostic and prognostic purposes, conclusions are inconsistent with regard to its relevance [89, 91-93]. Some studies associated low E-cadh expression with tumour size [93, 94], histological grade [93-95], distant metastasis and absence of oestrogen receptors, but not with lymph node status [94]. Although some investigators have found no association between E-cadh and the tumour stage, lymph node status or metastasis [89], other reported an association of E-cadh with lymph node status [93] and TNM stage [93].

With regard to prognosis, data is also controversial: many studies reported that E-cadh downregulation was associated with poor survival [80, 92-94, 96], but Gillett et al. [97] found that E-cadh reduction was a favorable prognostic factor, and Wang et al. [98] found no relationship between E-cadh and prognosis. These contrasting results may be associated to inter-study heterogeneity with respect to clinical data collection, immunohistochemical staining and interpretation as well as statistical modeling, that affect studies results [96].

Adhesion molecules have been extensively studied in canine mammary tumours [66-68, 70, 71, 83, 99-104]. In these tumours, E-cadh imunoexpression was first described by Restucci and coworkers [66], reporting low expression in malignant neoplasia. After, multiple studies showed reduced membranous expression of E-cadh in malignant mammary tumours, suggesting that down-regulation of adhesion molecules is a common event in canine mammary tumours [68, 71, 100]. Some studies associated low membrane expression of E-cadh and the histological type, poor differentiation, invasiveness, high proliferation and lymph node metastasis [66, 68, 71, 99, 100]. Reduction of E-cadh expression and associated catenins was also related with invasion and metastases on canine mammary tumours [66, 67, 70, 99]. These data hint a possible role of E-cadh in canine mammary tumours aggressiveness and on the emergence of an invasive and metastatic phenotype, suggesting E-cadh as a potential prognostic marker [67, 68, 71, 105]. However, although some studies found association between E-cadh and β-catenin expression and survival [68], other failed to find these associations [70, 100].

Regarding the feline mammary tumours, few studies have evaluated adhesion molecules expression and their prognostic value [106-108]. Loss of membrane E-cadh expression and its dislocation to the cytoplasm has been described in feline mammary carcinomas compared with benign tumours and hyperplastic or normal mammary tissues [106, 108, 109], as it was reported in canine mammary tumours [66, 68, 71, 100]. Albeit Penafiel-Verdu et al. [108] found an association between E-cadh expression and the grade and existence of regional metastasis at the moment of diagnosis, other studies fail to establish such associations [69,

109]. Abnormal cytoplasm E-cadh and/or catenin has been described in feline cancer cells [107-109] but not in the normal mammary tissue [107, 108], suggesting that this abnormal location might be related to a malignant transformation of feline mammary tumours [107]. Albeit not fully explored the prognostic role of E-cadh in feline mammary tumours, some studies could not establish an association between E-cadh and survival [109].

Understanding the metastatic progression of breast cancer in humans offers the opportunity to tailor individual based therapies by considering specific tumour characteristics [80]. For metastases to occur, several progressive changes are needed; these include neovascularization, decreased adherence of the tumour cells to each other, increased motility, adhesion to the extracellular matrix, and degradation of the extracellular matrix [80]. Downregulated expression of E-cadh destroys cell junctions and thus epithelial cells acquire the ability to migrate [93]. Consequently, decreased expression of E-cadh facilitates tumour invasion and metastasis [87, 88, 93]. Nevertheless, in the vast majority of human breast cancers, E-cadh expression is not lost, but retained [93]. Some studies in human reported preserved E-cadh expression in high aggressive breast cancer [110], such as the inflammatory breast cancer, in which E-cadh is not only retained but overexpressed [90, 95, 111] and distributed circumferentially 360° around the cancer cell membrane [93]. E-cadh accumulation and, subsequently, overexpression is responsible for the formation of lymphovascular emboli, conferring resistance to apoptosis and a survival advantage. It can be argued that, in the setting of the lymphovascular tumoral embolus, E-cadh is functioning not as a suppressor gene but rather as an oncogene [90]. More recently, Chu et al. [112] showed that E-cadh expression plays an important role in regulating tumourigenicity and hypoxia responses in an inflammatory breast cancer model: the loss of E-cadh and/or overexpression of its repressors, such as ZEB1, downregulated the expression of hypoxia-inducible 1α transcription factor (HIF-1α), leading to a reduction of the extracellular acidification of inflammatory breast cancer, tumour growth and metastasis formation [113].

Although multiple studies reported that metastases were more common in women with breast cancer with absent or low E-cadh expression [91, 114], recent studies have called into question whether cancer cells require the loss of E-cadh to invade and metastasize [89]. Besides, results on the expression of E-cadh in both primary tumour and lymph node metastasis are contradictory. Some studies reported lymph node metastasis with reduced or loss of E-cadh expression [115], whereas others reported overexpression [83, 116] or a similar expression to that observed in the primary tumour [83, 114]. E-cadh overexpression in canine mammary lymph node metastases was also reported by some studies [66, 70]. In the female cat, differences in E-cadh pattern of expression between primary tumors and regional metastases were described [69, 109], suggesting that during mammary carcinogenic progression, there is a dynamic and reversible modulation of the E-cadh complex [69]. E-cadh overexpression in metastases might be associated to their stabilization in the new environment, in order to adhere and to re-establish tissue architecture [116]. Although the mechanism behind this re-expression remains unclear [70], it has been suggested that E-cadh expression is not usually inactivated but its expression is dynamically modulated during the metastatic cascade. This plasticity is probably related to the flexibility of adhesion complexes, which might be temporarily downregulated or not expressed in primary tumours allowing cell detachment and invasion, and lately recovered in the metastatic site, thus favoring the survival and growth of metastatic cells [66, 116]. So, the loss of E-cadh expression might be a transient phenomenon that allows malignant cells to invade vascular channels and tissues; once in circulation, cancer cells re-express E-cadh, facilitating intercellular adhesion and enabling the formation of cohesive tumour emboli [95].

As described previously, cancer invasion and metastasis are, in fact, highly versatile processes, regulated at multiple levels. Recent studies indicate that cancer cells utilize two major migratory strategies: preserving intercellular cohesion as a collective, or as single-cell invasion into the surrounding stroma. The process of cancer cell individualization and acquisition of an invasive migratory phenotype commonly occurs within the framework of EMT. During EMT, epithelial cells undergo major transcriptional and morphological transformations, resulting in the loss of their intercellular adhesions, and the acquisition of mesenchymal-like properties [117]. As EMT progresses, the transformed cells lose the junctional connections with their neighbours, disengage from the epithelial layer in which they originated, and express mesenchymal markers such as N-cadh, vimentin, and a multitude of specific transcription factors (e.g., Snail, Slug, Twist) [117, 118]. The acquired mesenchymal phenotype is manifested in enhanced migratory activity, extracellular matrix production, invasiveness, and elevated resistance to apoptosis [118, 119]. These changes enable the cells to enter into small vessels and disseminate to distant organs, where they form metastasis [117].

There is controversy about the corollary of the type of E-cadh inactivation (gene mutation or promoter hypermethylation/repression) and the aggressiveness of tumour cells [87]. The infiltrating lobular carcinoma of human breast is classically characterized by a loss of E-cadh immunostaining and intercellular cohesion that results from truncating mutations or epigenetic bi-allelic silencing of E-cadh gene [89, 120-122].

In some breast carcinomas, invasion and metastasis is promoted when E-cadh expression is lost by the promoter methylation or repression by Snail/Slug and other EMT mediators [87, 90]. EMT has been associated with the metastatic cascade in several types of carcinomas, including human breast carcinomas [123, 124]; Lombaerts et al. [87] suggest that E-cadh promoter methylation, but not mutational inactivation, is part of the EMT programme, resulting in increased invasiveness and tumourigenic capacity in breast cancer. The molecular events of this programme can be inferred from the differentially expressed genes and include genes from the TGFβ pathway, transcription factors involved in E-cadh regulation (i.e., ZFHX1B, SNAI2, but not SNAI1, TWIST), annexins, AP1/2 transcription factors and members of the actin and intermediate filament cytoskeleton organization. Altered expression of these transcription factors seems to be also associated with an altered overexpression of transcriptional repressors of E-cadh in tumour cells; thus, considering that metastasis is facilitated by

EMT, the disturbance of this process might prevent breast cancer dissemination [87].

Recently, Raposo-Ferreira et al. [125] provided evidences that EMT plays an important role in the metastatic process of canine mammary carcinomas, describing a significantly high co-expression of E-Cadh$^+$/vimentin$^+$ in primary mammary carcinomas, especially in high grade carcinomas, when compared to their paired metastases. This distinct expression pattern suggests that EMT is also a dynamic reversible process in canine mammary tumours [125]. In fact, accumulating evidence supports a phenotypic plasticity of metastatic cells that allows a reverse process at a secondary site, known as mesenchymal-epithelial transition (MET), that promotes a transition back into an epithelial phenotype which will allow secondary tumour growth [27].

A few studies investigated the expression of EMT-inducing transcription factors, such as Snail, Slug or Zeb, and their correlation with E-cadh in canine or feline mammary neoplasia. With regard to Snail, Im et al. [126] found no association between Snail and E-cadh expression, although Snail expression in canine mammary tumours was significantly correlated with aggressive clinicopathological features such as histological grade and lymphatic invasion. In contrast, Gamba et al. [127] reported a direct association between E-cadh downregulation and Snail up-regulation in canine invasive micropapillary mammary carcinoma; however, no significant correlation was found between E-cadh and Zeb2 in this histological type [128]. Pang et al. [129] demonstrated that EMT induction by TGFβ can enrich cancer cells with stem cell properties.

Classic EMT is usually defined by morphological changes combined with the loss of E-cadh. However, *in vitro* studies using human breast cancer cell lines showed that EMT might be possible without E-cadh loss [130]. Timmermans-Sprang and collaborators [131] described that P-cadh mutations are associated with an EMT phenotype in canine mammary cell lines with E-cadh expression. In fact, in human breast cancer, the presence of P-cadh in an E-cadh positive background can promote invasion [132].

Regarding feline mammary tumours, Buendia et al. [133] found a negative association between E-cadh and N-cadh expression and reported an association between N-cadh and the tumour histological grade and regional metastasis; these authors suggested that N-cadh expression could be considered a sign of malignancy. Another study showed a co-expression of E-cadh and P-cadh in both primary and metastastic lesions and reported a large number of P-cadh positive feline mammary tumours suggesting that cadherins switching to P-cadh could play an important role in feline mammary tumourigenesis [134]. This hypothesis deserve additional investigation.

CONCLUSION

E-cadh is a classic cadherin that plays an important role in epithelial cell-to-cell adhesion, with recognized functions in morphogenetic processes and in the maintenance of normal tissue architecture. Alterations of E-cadh expression/function have been reported in physiological processes such as throughout the oestrous cycle or during the embryo-maternal interaction at embryo implantation. In contrast, E-cadh loss or reduced expression/function has also been associated with disease, namely with tumourigenesis and malignant progression of carcinomas. Breast cancer is one of the most studied cancer types regarding E-cadh relevance in the tumourigenesis, tumour progression and prognosis. Although the significant number of *in vitro* and *in vivo* investigations in this subject, several questions are still controversial. There is no consensus as to consider E-cadh as a prognostic marker, and its importance during tumour progression differs across the available literature. Nevertheless, E-cadh loss has long been described as a hallmark of EMT, a dynamic process that is especially relevant for breast cancer neoplastic invasion, progression and metastasis formation. In the veterinary setting, although fewer studies are available, especially for the endometrial cancer, results seem similar to the ones described for humans. Future studies are welcome, aiming the factual relevance of E-cadh in the metastatic cascade, focusing in the complex tumour-microenvironment interactions, both at the primary tumour and at the metastatic niche.

REFERENCES

[1] Colás-Algora, N., and J. Millán. 2019. "How many cadherins do human endothelial cells express?" *Cell Mol Life Sci* 76 (7):1299-1317. doi: 10.1007/s00018-018-2991-9.

[2] Pokutta, S., and W. I. Weis. 2007. "Structure and mechanism of cadherins and catenins in cell-cell contacts." *Annu Rev Cell Dev Biol* 23:237-61. doi: 10.1146/annurev.cellbio.22.010305.104241.

[3] Nollet, F., P. Kools, and F. van Roy. 2000. "Phylogenetic analysis of the cadherin superfamily allows identification of six major subfamilies besides several solitary members." *J Mol Biol* 299 (3):551-72. doi: 10.1006/jmbi.2000.3777.

[4] Oda, H., and M. Takeichi. 2011. "Evolution: structural and functional diversity of cadherin at the adherens junction." *J Cell Biol* 193 (7):1137-46. doi: 10.1083/jcb.201008173.

[5] Halbleib, J. M., and W. J. Nelson. 2006. "Cadherins in development: cell adhesion, sorting, and tissue morphogenesis." *Genes Dev* 20 (23):3199-214. doi: 10.1101/gad.1486806.

[6] Gumbiner, B. M. 2016. "Classical Cadherins." In *The Cadherin Superfamily: Key Regulators of Animal Development and Physiology*, edited by Shintaro T. Suzuki and Shinji Hirano, 41-69. Tokyo: Springer Japan.

[7] Garcia, M. A., W. J. Nelson, and N. Chavez. 2018. "Cell-Cell Junctions Organize Structural and Signaling Networks." *Cold Spring Harb Perspect Biol* 10 (4). doi: 10.1101/cshperspect.a029181.

[8] Braga, V. 2016. "Spatial integration of E-cadherin adhesion, signalling and the epithelial cytoskeleton." *Curr Opin Cell Biol* 42:138-145. doi: 10.1016/j.ceb.2016.07.006.

[9] Gloushankova, N. A., S. N. Rubtsova, and I. Y. Zhitnyak. 2017. "Cadherin-mediated cell-cell interactions in normal and cancer cells." *Tissue Barriers* 5 (3):e1356900. doi: 10.1080/21688370.2017. 1356900.

[10] Brüser, L., and S. Bogdan. 2017. "Adherens Junctions on the Move-Membrane Trafficking of E-Cadherin." *Cold Spring Harb Perspect Biol* 9 (3). doi: 10.1101/cshperspect.a029140.

[11] Bajpai, A., J. Tong, W. Qian, Y. Peng, and W. Chen. 2019. "The Interplay Between Cell-Cell and Cell-Matrix Forces Regulates Cell Migration Dynamics." *Biophys J* 117 (10):1795-1804. doi: 10.1016/j.bpj.2019.10.015.

[12] Katsuno-Kambe, H., and A. S. Yap. 2020. "Endocytosis, cadherins and tissue dynamics." *Traffic* 21: 268– 273. doi: 10.1111/tra.12721.

[13] Takeichi, M. 2014. "Dynamic contacts: rearranging adherens junctions to drive epithelial remodelling." *Nat Rev Mol Cell Biol* 15 (6):397-410. doi: 10.1038/nrm3802.

[14] Mège, R. M., and N. Ishiyama. 2017. "Integration of Cadherin Adhesion and Cytoskeleton at." *Cold Spring Harb Perspect Biol* 9 (5). doi: 10.1101/cshperspect.a028738.

[15] Kowalczyk, A. P., and B. A. Nanes. 2012. "Adherens junction turnover: regulating adhesion through cadherin endocytosis, degradation, and recycling." *Subcell Biochem* 60:197-222. doi: 10.1007/978-94-007-4186-7_9.

[16] Nagathihalli, N. S., and N. B. Merchant. 2012. "Src-mediated regulation of E-cadherin and EMT in pancreatic cancer." *Front Biosci (Landmark Ed)* 17:2059-69. doi: 10.2741/4037.

[17] Park, J. I., S. W. Kim, J. P. Lyons, H. Ji, T. T. Nguyen, K. Cho, M. C. Barton, T. Deroo, K. Vleminckx, R. T. Moon, and P. D. McCrea. 2005. "Kaiso/p120-catenin and TCF/beta-catenin complexes coordinately regulate canonical Wnt gene targets." *Dev Cell* 8 (6):843-54. doi: 10.1016/j.devcel.2005.04.010.

[18] Bertocchi, C., M. Vaman Rao, and R. Zaidel-Bar. 2012. "Regulation of adherens junction dynamics by phosphorylation switches." *J Signal Transduct* 2012:125295. doi: 10.1155/2012/125295.

[19] Payan-Carreira, R., M. A. Pires, C. Santos, B. S. Holst, J. Colaço, and H. Rodriguez-Martinez. 2016. "Immunolocalization of E-cadherin and β-catenin in the cyclic and early pregnant canine

endometrium." *Theriogenology* 86 (4):1092-101. doi: 10.1016/j.theriogenology.2016.03.041.

[20] Huang, H., S. Wright, J. Zhang, and R. A. Brekken. 2019. "Getting a grip on adhesion: Cadherin switching and collagen signaling." *Biochim Biophys Acta Mol Cell Res* 1866 (11):118472. doi: 10.1016/j.bbamcr.2019.04.002.

[21] Xiao, X., Y. Ni, C. Yu, L. Li, B. Mao, Y. Yang, D. Zheng, B. Silvestrini, and C. Y. Cheng. 2018. "Src family kinases (SFKs) and cell polarity in the testis." *Semin Cell Dev Biol* 81:46-53. doi: 10.1016/j.semcdb.2017.11.024.

[22] Xiao, X., Y. Yang, B. Mao, C. Y. Cheng, and Y. Ni. 2019. "Emerging Role for SRC family kinases in junction dynamics during spermatogenesis." *Reproduction*. doi: 10.1530/REP-18-0440.

[23] Serrels, A., M. Canel, V. G. Brunton, and M. C. Frame. 2011. "Src/FAK-mediated regulation of E-cadherin as a mechanism for controlling collective cell movement: insights from in vivo imaging." *Cell Adh Migr* 5 (4):360-5. doi: 10.4161/cam.5.4.17290.

[24] Dosch, A. R., X. Dai, A. A. Gaidarski Iii, C. Shi, J. A. Castellanos, M. N. VanSaun, N. B. Merchant, and N. S. Nagathihalli. 2019. "Src kinase inhibition restores E-cadherin expression in dasatinib-sensitive pancreatic cancer cells." *Oncotarget* 10 (10):1056-1069. doi: 10.18632/oncotarget.26621.

[25] Knights, A. J., A. P. Funnell, M. Crossley, and R. C. Pearson. 2012. "Holding Tight: Cell Junctions and Cancer Spread." *Trends Cancer Res* 8:61-69.

[26] Bruner, H. C., and P. W. B. Derksen. 2018. "Loss of E-Cadherin-Dependent Cell-Cell Adhesion and the Development and Progression of Cancer." *Cold Spring Harb Perspect Biol* 10 (3). doi: 10.1101/cshperspect.a029330.

[27] Chaffer, C. L., B. P. San Juan, E. Lim, and R. A. Weinberg. 2016. "EMT, cell plasticity and metastasis." *Cancer Metastasis Rev* 35 (4):645-654. doi: 10.1007/s10555-016-9648-7.

[28] Nieto, M. A., R. Y. Huang, R. A. Jackson, and J. P. Thiery. 2016. "EMT: 2016." *Cell* 166 (1):21-45. doi: 10.1016/j.cell.2016.06.028.

[29] Pastushenko, I., A. Brisebarre, A. Sifrim, M. Fioramonti, T. Revenco, S. Boumahdi, A. Van Keymeulen, D. Brown, V. Moers, S. Lemaire, S. De Clercq, E. Minguijon, C. Balsat, Y. Sokolow, C. Dubois, F. De Cock, S. Scozzaro, F. Sopena, A. Lanas, N. D'Haene, I. Salmon, J. C. Marine, T. Voet, P. A. Sotiropoulou, and C. Blanpain. 2018. "Identification of the tumour transition states occurring during EMT." *Nature* 556 (7702):463-468. doi: 10.1038/s41586-018-0040-3.

[30] Petrova, Y. I., L. Schecterson, and B. M. Gumbiner. 2016. "Roles for E-cadherin cell surface regulation in cancer." *Mol Biol Cell* 27 (21):3233-3244. doi: 10.1091/mbc.E16-01-0058.

[31] Venhuizen, J. H., F. J. C. Jacobs, P. N. Span, and M. M. Zegers. 2019. "P120 and E-cadherin: Double-edged swords in tumor metastasis." *Seminars in Cancer Biology*. doi: https://doi.org/10.1016/j.semcancer.2019.07.020.

[32] Loh, C. Y., J. Y. Chai, T. F. Tang, W. F. Wong, G. Sethi, M. K. Shanmugam, P. P. Chong, and C. Y. Looi. 2019. "The E-Cadherin and N-Cadherin Switch in Epithelial-to-Mesenchymal Transition: Signaling, Therapeutic Implications, and Challenges." *Cells* 8 (10). doi: 10.3390/cells8101118.

[33] Ferreira, A. C., G. Suriano, N. Mendes, B. Gomes, X. Wen, F. Carneiro, R. Seruca, and J. C. Machado. 2012. "E-cadherin impairment increases cell survival through Notch-dependent upregulation of Bcl-2." *Hum Mol Genet* 21 (2):334-43. doi: 10.1093/hmg/ddr469.

[34] Rodriguez, F. J., L. J. Lewis-Tuffin, and P. Z. Anastasiadis. 2012. "E-cadherin's dark side: Possible role in tumor progression." *Biochimica et Biophysica Acta (BBA) - Reviews on Cancer* 1826 (1):23-31. doi: https://doi.org/10.1016/j.bbcan.2012.03.002.

[35] Gallegos, L. L., and J. S. Brugge. 2014. "Live free or die: cell-cell adhesion regulates sensitivity to trail-induced apoptosis." *Dev Cell* 30 (1):3-4. doi: 10.1016/j.devcel.2014.06.031.

[36] Lu, M., S. Marsters, X. Ye, E. Luis, L. Gonzalez, and A. Ashkenazi. 2014. "E-cadherin couples death receptors to the cytoskeleton to

regulate apoptosis." *Mol Cell* 54 (6):987-98. doi: 10.1016/j.molcel.2014.04.029.

[37] Capra, J., and S. Eskelinen. 2017. "Correlation between E-cadherin interactions, survivin expression, and apoptosis in MDCK and ts-Src MDCK cell culture models." *Lab Invest* 97 (12):1453-1470. doi: 10.1038/labinvest.2017.89.

[38] Carico, E., M. Atlante, E. Giarnieri, S. Raffa, B. Bucci, M. R. Giovagnoli, and A. Vecchione. 2010. "E-cadherin and alpha-catenin expression in normal, hyperplastic and neoplastic endometrium." *Anticancer Res* 30 (12):4993-7.

[39] Rahnama, F., B. Thompson, M. Steiner, F. Shafiei, P. E. Lobie, and M. D. Mitchell. 2009. "Epigenetic regulation of E-cadherin controls endometrial receptivity." *Endocrinology* 150 (3):1466-1472.

[40] Horne, A. W., J. O. White, and el-N Lalani. 2002. "Adhesion molecules and the normal endometrium." *BJOG* 109 (6):610-7. doi: 10.1111/j.1471-0528.2002.t01-1-01017.x.

[41] van der Linden, P. J., A. F. de Goeij, G. A. Dunselman, H. W. Erkens, and J. L. Evers. 1995. "Expression of cadherins and integrins in human endometrium throughout the menstrual cycle." *Fertil Steril* 63 (6):1210-6.

[42] Fujimoto, J., S. Ichigo, M. Hori, and T. Tamaya. 1996. "Alteration of E-cadherin, alpha- and beta-catenin mRNA expression in human uterine endometrium during the menstrual cycle." *Gynecol Endocrinol* 10 (3):187-91.

[43] Matsuzaki, S., C. Darcha, E. Maleysson, M. Canis, and G. Mage. 2010. "Impaired down-regulation of E-cadherin and beta-catenin protein expression in endometrial epithelial cells in the mid-secretory endometrium of infertile patients with endometriosis." *J Clin Endocrinol Metab* 95 (7):3437-45. doi: 10.1210/jc.2009-2713.

[44] Satterfield, M. C., K. A. Dunlap, K. Hayashi, R. C. Burghardt, T. E. Spencer, and F. W. Bazer. 2007. "Tight and adherens junctions in the ovine uterus: differential regulation by pregnancy and progesterone." *Endocrinology* 148 (8):3922-31. doi: 10.1210/en.2007-0321.

[45] Ryan, P. L., D. L. Baum, J. A. Lenhart, K. M. Ohleth, and C. A. Bagnell. 2001. "Expression of uterine and cervical epithelial cadherin during relaxin-induced growth in pigs." *Reproduction* 122 (6):929-37.

[46] Guo, B., B. C. Han, Z. Tian, X. M. Zhang, L. X. Jiang, J. X. Liu, and Z. P. Yue. 2009. "Expression and Hormonal Regulation of E-Cadherin in Canine Uterus During Early Pregnancy." *Reprod Domest Anim*. doi: 10.1111/j.1439-0531.2009.01550.x.

[47] Shih, I. M., M. Y Hsu, R. J. Oldt, M. Herlyn, J. D. Gearhart, and R. J. Kurman. 2002. "The Role of E-cadherin in the Motility and Invasion of Implantation Site Intermediate Trophoblast." *Placenta* 23 (10):706-715. doi: https://doi.org/10.1053/plac.2002.0864.

[48] Kiewisz, J., M. M. Kaczmarek, A. Andronowska, A. Blitek, and A. J. Ziecik. 2011. "Gene expression of WNTs, β-catenin and E-cadherin during the periimplantation period of pregnancy in pigs--involvement of steroid hormones." *Theriogenology* 76 (4):687-99. doi: 10.1016/j.theriogenology.2011.03.022.

[49] Dudley, J. S., C. R. Murphy, M. B. Thompson, T. Carter, and B. M. McAllan. 2018. "Uterine Epithelial Cells Undergo a Plasma Membrane Transformation During Early Pregnancy in the Domestic Cat (Felis catus)." *The Anatomical Record* 301 (9):1497-1505. doi: 10.1002/ar.23895.

[50] Sakuragi, N., M. Nishiya, K. Ikeda, T. Ohkouch, E. E. Furth, H. Hareyama, C. Satoh, and S. Fujimoto. 1994. "Decreased E-cadherin expression in endometrial carcinoma is associated with tumor dedifferentiation and deep myometrial invasion." *Gynecol Oncol* 53 (2):183-9. doi: 10.1006/gyno.1994.1113.

[51] Siitonen, S. M., J. T. Kononen, H. J. Helin, I. S. Rantala, K. A. Holli, and J. J. Isola. 1996. "Reduced E-cadherin expression is associated with invasiveness and unfavorable prognosis in breast cancer." *Am J Clin Pathol* 105 (4):394-402. doi: 10.1093/ajcp/105.4.394.

[52] Koyuncuoglu, M., E. Okyay, B. Saatli, S. Olgan, M. Akin, and U. Saygili. 2012. "Tumor budding and E-Cadherin expression in endometrial carcinoma: are they prognostic factors in endometrial

cancer?" *Gynecol Oncol* 125 (1):208-13. doi: 10.1016/j.ygyno. 2011.12.433.

[53] Yalta, T., L. Atay, F. Atalay, M. Caydere, M. Gonultas, and H. Ustun. 2009. "E-cadherin expression in endometrial malignancies: comparison between endometrioid and non-endometrioid carcinomas." *J Int Med Res* 37 (1):163-8. doi: 10.1177/147323 000903700119.

[54] Mell, L. K., J. J. Meyer, M. Tretiakova, A. Khramtsov, C. Gong, S. D. Yamada, A. G. Montag, and A. J. Mundt. 2004. "Prognostic significance of E-cadherin protein expression in pathological stage I-III endometrial cancer." *Clin Cancer Res* 10 (16):5546-53. doi: 10.1158/1078-0432.CCR-0943-03.

[55] González-Rodilla, I., L. Aller, J. Llorca, A. B. Muñoz, V. Verna, J. Estévez, and J. Schneider. 2013. "The E-Cadherin expression vs. tumor cell proliferation paradox in endometrial cancer." *Anticancer Res* 33 (11):5091-5.

[56] Sugihara, T. 2016. "Loss of Adherens Junction Protein E-Cadherin is a Biomarker of High- Grade Histology and Poor Prognosis in Endometrial Cancer | Insight Medical Publishing." *Annals of Clinical and Laboratory Research* 4. doi: 10.21767/2386-5180.100055.

[57] Schlosshauer, P. W., L. H. Ellenson, and R. A. Soslow. 2002. "Beta-catenin and E-cadherin expression patterns in high-grade endometrial carcinoma are associated with histological subtype." *Mod Pathol* 15 (10):1032-7. doi: 10.1097/01.MP.0000028573.34289.04.

[58] Gil da Costa, R. M., M. Santos, I. Amorim, C. Lopes, P. D. Pereira, and A. M. Faustino. 2009. "An immunohistochemical study of feline endometrial adenocarcinoma." *J Comp Pathol* 140 (4):254-9. doi: 10.1016/j.jcpa.2008.12.006.

[59] Saraiva, A. L., Payan-Carreira, R., Gärtner, F., Pires, M. A. 2012. "Feline Endometrial Adenocarcinomas." In *Adenocarcinoma: Pathogenesis, Treatment and Prognosis*, edited by M. A. Alcalá Longoria, J. I., 175-189. Nova Science Publishers Inc.

[60] Payan-Carreira, R., A. L. Saraiva, T. Santos, H. Vilhena, A. Sousa, C. Santos, and M. A. Pires. 2013. "Feline Endometrial

Adenocarcinoma in Females < 1 Year Old: A Description of Four Cases." *Reproduction in Domestic Animals* 48 (5):e70-e77. doi: 10.1111/rda.12190.

[61] Sontas, B. H., Ö Erdogan, S. Ö Apaydin Enginler, Ö Turna Yilmaz, G. Şennazli, and H. Ekici. 2013. "Endometrial adenocarcinoma in two young queens." *Journal of Small Animal Practice* 54 (3):156-159. doi: 10.1111/j.1748-5827.2012.01307.x.

[62] Sapierzynski, R. A. Dolka, I. Cywinska, A. 2009. "Multiple pathologies of the feline uterus: a case report." *Veterinarni Medicina* 52 (7):345–350.

[63] Pires, M. A., A. L. Saraiva, H. Vilhena, S. Miranda, I. Fonseca, P. Moreira, A. M. Alves, R. Paiva, and R. Payan-Carreira. 2013. "Endometrial adenocarcinoma in a cat with abdominal mestastasis." *Journal of Comparative Pathology* 148 (1):67-67.

[64] Saraiva, A. L., R. Payan-Carreira, F. Gärtner, M. R. Fortuna da Cunha, A. Rêma, F. Faria, L. M. Lourenço, and M.os A Pires. 2015. "An immunohistochemical study on the expression of sex steroid receptors, Ki-67 and cytokeratins 7 and 20 in feline endometrial adenocarcinomas." *BMC Vet Res* 11:204. doi: 10.1186/s12917-015-0530-6.

[65] Pires, M. A., F. Seixas, C. Palmeira, and R. Payan-Carreira. 2010. "Histopathologic and Immunohistochemical Exam in One Case of Canine Endometrial Adenocarcinoma." *Reproduction in Domestic Animals* 45 (3):545-549. doi: 10.1111/j.1439-0531.2008.01243.x.

[66] Restucci, B., S. Papparella, G. De Vico, and P. Maiolino. 1997. "E cadherin expression in normal and neoplastic canine mammary gland." *Journal of Comparative Pathology* 116 (2):191-202. doi: https://doi.org/10.1016/S0021-9975(97)80076-X.

[67] Sarli, G., R. Preziosi, L. De Tolla, B. Brunetti, and C. Benazzi. 2004. "E-cadherin immunoreactivity in canine mammary tumors." *J Vet Diagn Invest* 16 (6):542-7. doi: 10.1177/104063870401600608.

[68] Gama, A., J. Paredes, F. Gartner, A. Alves, and F. Schmitt. 2008. "Expression of E-cadherin, P-cadherin and beta-catenin in canine malignant mammary tumours in relation to clinicopathological

parameters, proliferation and survival." *Vet J* 177 (1):45-53. doi: 10.1016/j.tvjl.2007.05.024.

[69] Figueira, A. C., C. Gomes, J. Tavares de Oliveira, H. Vilhena, J. Carvalheira, A. J. F. de Matos, P. Dias Pereira, and F. Gaertner. 2014. "Aberrant P-cadherin expression is associated to aggressive feline mammary carcinomas." *Bmc Veterinary Research* 10. doi: 10.1186/s12917-014-0270-z.

[70] Brunetti, B., G. Sarli, R. Preziosi, S. Leprotti, and C. Benazzi. 2003. "E-cadherin expression in canine mammary carcinomas with regional lymph node metastases." *J Vet Med A Physiol Pathol Clin Med* 50 (10):496-500. doi: 10.1111/j.1439-0442.2003.00577.x.

[71] De Matos, A. J., C. C. Lopes, A. M. Faustino, J. G. Carvalheira, G. R. Rutteman, and M.e F Gärtner. 2007. "E-cadherin, beta-catenin, invasion and lymph node metastases in canine malignant mammary tumours." *APMIS* 115 (4):327-34. doi: 10.1111/j.1600-0463.2007.apm_544.x.

[72] Park, D., R. KÅResen, U. Axcrona, T. Noren, and T. Sauer. 2007. "Expression pattern of adhesion molecules (E-cadherin, α-, β-, γ-catenin and claudin-7), their influence on survival in primary breast carcinoma, and their corresponding axillary lymph node metastasis." *APMIS* 115 (1):52-65. doi: 10.1111/j.1600-0463.2007.apm_524.x.

[73] Figueira, A. C. Gomes, C. Vilhena, H. Miranda, S. Carvalheira, J De Matos, A. J. Dias-Pereira, P. Gärtner F. 2015. "Characterization of alpha-, beta- and p120-Catenin Expression in Feline Mammary Tissues and their Relation with E- and P-Cadherin." *Anticancer Res* 35 (6):3361-3369.

[74] Palacios, J., N. Benito, A. Pizarro, A. Suarez, J. Espada, A. Cano, and C. Gamallo. 1995. "Anomalous expression of P-cadherin in breast carcinoma. Correlation with E-cadherin expression and pathological features." *Am J Pathol* 146 (3):605-612.

[75] Gama, A., J. Paredes, M. F. Milanezi, J. S. Reis-Filho, F. Gartner, and F. C. Schmitt. 2002. "P-cadherin expression in canine lactating mammary gland." *J Cell Biochem* 86 (3):420-421. doi: 10.1002/jcb.10245.

[76] Gama, A., J. Paredes, A. Albergaria, F. Gartner, and F. Schmitt. 2004. "P-cadherin expression in canine mammary tissues." *J Comp Pathol* 130 (1):13-20. doi: 10.1016/s0021-9975(03)00064-1.

[77] Visvader, J. E., and J. Stingl. 2014. "Mammary stem cells and the differentiation hierarchy: current status and perspectives." *Genes Dev* 28 (11):1143-1158. doi: 10.1101/gad.242511.114.

[78] Derksen, P. W., T. M. Braumuller, E. van der Burg, M. Hornsveld, E. Mesman, J. Wesseling, P. Krimpenfort, and J. Jonkers. 2011. "Mammary-specific inactivation of E-cadherin and p53 impairs functional gland development and leads to pleomorphic invasive lobular carcinoma in mice." *Dis Model Mech* 4 (3):347-358. doi: 10.1242/dmm.006395.

[79] Xi, Y., X. Zhang, Z. Yang, Q. Guo, Z. Zhang, S. Chen, H. Zheng, and B. Hua. 2019. "Prognositic significance of P-cadherin expression in breast cancer: Protocol for a meta-analysis." *Medicine (Baltimore)* 98 (12):e14924. doi: 10.1097/MD.0000000000014924.

[80] Heimann, R., F. Lan, R. McBride, and S. Hellman. 2000. "Separating favorable from unfavorable prognostic markers in breast cancer: the role of E-cadherin." *Cancer Res* 60 (2):298-304.

[81] Siegel, R. L., K. D. Miller, and A. Jemal. 2018. "Cancer statistics, 2018." *CA Cancer J Clin* 68 (1):7-30. doi: 10.3322/caac.21442.

[82] Misdorp, W. 2002. "Tumors of the Mammary Gland." *Tumors in Domestic Animals*:575-606. doi: doi:10.1002/9780470376928.ch12.

[83] Gama, A., and F. Schmitt. 2012. "Cadherin cell adhesion system in canine mammary cancer: a review." *Vet Med Int* 2012:357187. doi: 10.1155/2012/357187.

[84] Rutteman, G. R., S. J. Withrow, and E. G. MacEwenm. 2001. "Tumours of the mammary gland." In *Small Animal Clinical Oncology*, edited by SJ Withrow and EG MacEwen, 455-477. Philadelphia: W. B. Saunders Comp.

[85] Millanta, F., G. Lazzeri, I. Vannozzi, P. Viacava, and A. Poli. 2002. "Correlation of vascular endothelial growth factor expression to overall survival in feline invasive mammary carcinomas." *Vet Pathol* 39 (6):690-696. doi: 10.1354/vp.39-6-690.

[86] Seixas, F., C. Palmeira, M. A. Pires, M. J. Bento, and C. Lopes. 2011. "Grade is an independent prognostic factor for feline mammary carcinomas: a clinicopathological and survival analysis." *Vet J* 187 (1):65-71. doi: 10.1016/j.tvjl.2009.10.030.

[87] Lombaerts, M., T. van Wezel, K. Philippo, J. W. Dierssen, R. M. Zimmerman, J. Oosting, R. van Eijk, P. H. Eilers, B. van de Water, C. J. Cornelisse, and A. M. Cleton-Jansen. 2006. "E-cadherin transcriptional downregulation by promoter methylation but not mutation is related to epithelial-to-mesenchymal transition in breast cancer cell lines." *Br J Cancer* 94 (5):661-71. doi: 10.1038/sj.bjc.6602996.

[88] Horne, H. N., H. Oh, M. E. Sherman, M. Palakal, S. M. Hewitt, M. K. Schmidt, R. L. Milne, D. Hardisson, J. Benitez, C. Blomqvist, M. K. Bolla, H. Brenner, J. Chang-Claude, R. Cora, F. J. Couch, K. Cuk, P. Devilee, D. F. Easton, D. M. Eccles, U. Eilber, J. M. Hartikainen, P. Heikkilä, B. Holleczek, M. J. Hooning, M. Jones, R. Keeman, A. Mannermaa, J. W. M. Martens, T. A. Muranen, H. Nevanlinna, J. E. Olson, N. Orr, J. I. A. Perez, P. D. P. Pharoah, K. J. Ruddy, K. U. Saum, M. J. Schoemaker, C. Seynaeve, R. Sironen, V. T. H.BM Smit, A. J. Swerdlow, M. Tengström, A. S. Thomas, A. M. Timmermans, R. A. E.M Tollenaar, M. A. Troester, C. J. van Asperen, C. H. M. van Deurzen, F. F. Van Leeuwen, L. J. Van't Veer, M. García-Closas, and J. D. Figueroa. 2018. "E-cadherin breast tumor expression, risk factors and survival: Pooled analysis of 5,933 cases from 12 studies in the Breast Cancer Association Consortium." *Sci Rep* 8 (1):6574. doi: 10.1038/s41598-018-23733-4.

[89] Borcherding, N., K. Cole, P. Kluz, M. Jorgensen, R. Kolb, A. Bellizzi, and W. Zhang. 2018. "Re-Evaluating E-Cadherin and β-Catenin: A Pan-Cancer Proteomic Approach with an Emphasis on Breast Cancer." *Am J Pathol* 188 (8):1910-1920. doi: 10.1016/j.ajpath.2018.05.003.

[90] Ye, Y. Tellez, J. D. Durazo, M. Belcher, M. Yearsley, K. Barsky, S. H. 2010. "E-Cadherin Accumulation within the Lymphovascular

Embolus of Inflammatory Breast Cancer Is Due to Altered Trafficking." *Anticancer Research* 30:3903-3910.

[91] Gamallo, C., J. Palacios, A. Suarez, A. Pizarro, P. Navarro, M. Quintanilla, and A. Cano. 1993. "Correlation of E-cadherin expression with differentiation grade and histological type in breast carcinoma." *Am J Pathol* 142 (4):987-93.

[92] Kovács, A., J. Dhillon, and R. A. Walker. 2003. "Expression of P-cadherin, but not E-cadherin or N-cadherin, relates to pathological and functional differentiation of breast carcinomas." *Mol Pathol* 56 (6):318-22. doi: 10.1136/mp.56.6.318.

[93] Li, Z., S. Yin, L. Zhang, W. Liu, and B. Chen. 2017. "Prognostic value of reduced E-cadherin expression in breast cancer: a meta-analysis." *Oncotarget* 8 (10):16445-16455. doi: 10.18632/oncotarget.14860.

[94] Rakha, E. A., D. Abd El Rehim, S. E. Pinder, S. A. Lewis, and I. O. Ellis. 2005. "E-cadherin expression in invasive non-lobular carcinoma of the breast and its prognostic significance." *Histopathology* 46 (6):685-93. doi: 10.1111/j.1365-2559.2005.02156.x.

[95] Charafe-Jauffret, E., C. Tarpin, V. J. Bardou, F. Bertucci, C. Ginestier, A. C. Braud, B. Puig, J. Geneix, J. Hassoun, D. Birnbaum, J. Jacquemier, and P. Viens. 2004. "Immunophenotypic analysis of inflammatory breast cancers: identification of an 'inflammatory signature'." *J Pathol* 202 (3):265-73. doi: 10.1002/path.1515.

[96] Gould Rothberg, B. E., and M. B. Bracken. 2006. "E-cadherin immunohistochemical expression as a prognostic factor in infiltrating ductal carcinoma of the breast: a systematic review and meta-analysis." *Breast Cancer Res Treat* 100 (2):139-48. doi: 10.1007/s10549-006-9248-2.

[97] Gillett, C. E., D. W. Miles, K. Ryder, D. Skilton, R. D. Liebman, R. J. Springall, D. M. Barnes, and A. M. Hanby. 2001. "Retention of the expression of E-cadherin and catenins is associated with shorter survival in grade III ductal carcinoma of the breast." *J Pathol* 193 (4):433-41. doi: 10.1002/path.831.

[98] Wang, D., L. Su, D. Huang, H. Zhang, D. M. Shin, and Z. G. Chen. 2011. "Downregulation of E-Cadherin enhances proliferation of head and neck cancer through transcriptional regulation of EGFR." *Mol Cancer* 10:116. doi: 10.1186/1476-4598-10-116.

[99] Reis, A. L., J. Carvalheira, F. C. Schmitt, and F. Gartner. 2003. "Immunohistochemical study of the expression of E-cadherin in canine mammary tumours." *Vet Rec* 152 (20):621-624. doi: 10.1136/vr.152.20.621.

[100] Brunetti, B., G. Sarli, R. Preziosi, I. Monari, and C. Benazzi. 2005. "E-cadherin and beta-catenin reduction influence invasion but not proliferation and survival in canine malignant mammary tumors." *Vet Pathol* 42 (6):781-787. doi: 10.1354/vp.42-6-781.

[101] Torres, L. N., J. M. Matera, C. H. Vasconcellos, J. L. Avanzo, F. J. Hernandez-Blazquez, and M. L. Dagli. 2005. "Expression of connexins 26 and 43 in canine hyperplastic and neoplastic mammary glands." *Vet Pathol* 42 (5):633-641. doi: 10.1354/vp.42-5-633.

[102] Nowak, M., J. A. Madej, and P. Dziegiel. 2007. "Expression of E-cadherin, beta-catenin and Ki-67 antigen and their reciprocal relationships in mammary adenocarcinomas in bitches." *Folia Histochem Cytobiol* 45 (3):233-238.

[103] Nowak, M., J. A. Madej, M. Podhorska-Okolow, and P. Dziegiel. 2008. "Expression of extracellular matrix metalloproteinase (MMP-9), E-cadherin and proliferation-associated antigen Ki-67 and their reciprocal correlation in canine mammary adenocarcinomas." *In Vivo* 22 (4):463-469.

[104] Rodo, A., and E. Malicka. 2008. "E-cadherin immunohistochemical expression in mammary gland neoplasms in bitches." *Pol J Vet Sci* 11 (1):47-54.

[105] Yoshida, K., S. Yoshida, N. Choisunirachon, T. Saito, K. Matsumoto, K. Saeki, M. Mochizuki, R. Nishimura, N. Sasaki, and T. Nakagawa. 2014. "The relationship between clinicopathological features and expression of epithelial and mesenchymal markers in spontaneous canine mammary gland tumors." *J Vet Med Sci* 76 (10):1321-1327. doi: 10.1292/jvms.14-0104.

[106] Dias-Pereira, P., and F. Gärtner. 2003. "Expression of E-cadherin in normal, hyperplastic and neoplastic feline mammary tissue." *Vet Rec* 153 (10):297-302.

[107] Takauji, S. R., M. Watanabe, R. Uyama, T. Nakagawa, N. Miyajima, M. Mochizuki, R. Nishimura, S. Sugano, and N. Sasaki. 2007. "Expression and subcellular localization of E-cadherin, alpha-catenin, and beta-catenin in 8 feline mammary tumor cell lines." *J Vet Med Sci* 69 (8):831-834. doi: 10.1292/jvms.69.831.

[108] Penafiel-Verdu, C., A. J. Buendia, J. A. Navarro, G. A. Ramirez, M. Vilafranca, J. Altimira, and J. Sanchez. 2012. "Reduced expression of E-cadherin and beta-catenin and high expression of basal cytokeratins in feline mammary carcinomas with regional metastasis." *Vet Pathol* 49 (6):979-987. doi: 10.1177/ 0300985812436744.

[109] Zappulli, V., S. De Cecco, D. Trez, D. Caliari, L. Aresu, and M. Castagnaro. 2012. "Immunohistochemical expression of E-cadherin and beta-catenin in feline mammary tumours." *J Comp Pathol* 147 (2-3):161-170. doi: 10.1016/j.jcpa.2012.02.004.

[110] Howard, E. M., S. K. Lau, R. H. Lyles, G. G. Birdsong, J. N. Umbreit, and R. Kochhar. 2005. "Expression of e-cadherin in high-risk breast cancer." *J Cancer Res Clin Oncol* 131 (1):14-8. doi: 10.1007/s00432-004-0618-z.

[111] Kleer, C. G., K. L. van Golen, T. Braun, and S. D. Merajver. 2001. "Persistent E-cadherin expression in inflammatory breast cancer." *Mod Pathol* 14 (5):458-64. doi: 10.1038/modpathol.3880334.

[112] Chu, K., K. M. Boley, R. Moraes, S. H. Barsky, and F. M. Robertson. 2013. "The paradox of E-cadherin: role in response to hypoxia in the tumor microenvironment and regulation of energy metabolism." *Oncotarget* 4 (3):446-462. doi: 10.18632/oncotarget.872.

[113] Sousa, B., J. Pereira, and J. Paredes. 2019. "The Crosstalk Between Cell Adhesion and Cancer Metabolism." *Int J Mol Sci* 20 (8). doi: 10.3390/ijms20081933.

[114] Yoshida, R. Kimura, N. Harada, Y. Ohuchi, N. 2001. "The loss of E-cadherin, α- and β-catenin expression is associated with metastasis

and poor prognosis in invasive breast cancer." *International Journal of Oncology* 18 (13):513-520. doi: 10.3892/ijo.18.3.513.

[115] Yang, L., X. W. Wang, L. P. Zhu, H. L. Wang, B. Wang, Q. Zhao, and X. Y. Wang. 2018. "Significance and prognosis of epithelial-cadherin expression in invasive breast carcinoma." In *Oncol Lett*, 1659-1665.

[116] Bukholm, I. K., J. M. Nesland, and A. L. Børresen-Dale. 2000. "Re-expression of E-cadherin, α-catenin and β-catenin, but not of γ-catenin, in metastatic tissue from breast cancer patients." *The Journal of Pathology* 190 (1):15-19. doi: 10.1002/(SICI)1096-9896(200001) 190:1<15::AID-PATH489>3.0.CO;2-L.

[117] Elisha, Y., V. Kalchenko, Y. Kuznetsov, and B. Geiger. 2018. "Dual role of E-cadherin in the regulation of invasive collective migration of mammary carcinoma cells." *Sci Rep* 8 (1):4986. doi: 10.1038/s41598-018-22940-3.

[118] Kalluri, R., and R. A. Weinberg. 2009. "The basics of epithelial-mesenchymal transition." *J Clin Invest* 119 (6):1420-1428. doi: 10.1172/jci39104.

[119] Thiery, J. P., H. Acloque, R. Y. Huang, and M. A. Nieto. 2009. "Epithelial-mesenchymal transitions in development and disease." *Cell* 139 (5):871-890. doi: 10.1016/j.cell.2009.11.007.

[120] Berx, G., A. M. Cleton-Jansen, K. Strumane, W. J. de Leeuw, F. Nollet, F. van Roy, and C. Cornelisse. 1996. "E-cadherin is inactivated in a majority of invasive human lobular breast cancers by truncation mutations throughout its extracellular domain." *Oncogene* 13 (9):1919-1925.

[121] De Leeuw, W. J., G. Berx, C. B. Vos, J. L. Peterse, M. J. Van de Vijver, S. Litvinov, F. Van Roy, C. J. Cornelisse, and A. M. Cleton-Jansen. 1997. "Simultaneous loss of E-cadherin and catenins in invasive lobular breast cancer and lobular carcinoma in situ." *J Pathol* 183 (4):404-411. doi: 10.1002/(sici)1096-9896(199712) 183:4<404::aid-path1148>3.0.co;2-9.

[122] Sarrio, D., G. Moreno-Bueno, D. Hardisson, C. Sanchez-Estevez, M. Guo, J. G. Herman, C. Gamallo, M. Esteller, and J. Palacios. 2003.

"Epigenetic and genetic alterations of APC and CDH1 genes in lobular breast cancer: relationships with abnormal E-cadherin and catenin expression and microsatellite instability." *Int J Cancer* 106 (2):208-215. doi: 10.1002/ijc.11197.

[123] Tsuji, T., S. Ibaragi, and G. F. Hu. 2009. "Epithelial-mesenchymal transition and cell cooperativity in metastasis." *Cancer Res* 69 (18):7135-7139. doi: 10.1158/0008-5472.can-09-1618.

[124] Drasin, D. J., T. P. Robin, and H. L. Ford. 2011. "Breast cancer epithelial-to-mesenchymal transition: examining the functional consequences of plasticity." *Breast Cancer Res* 13 (6):226. doi: 10.1186/bcr3037.

[125] Raposo-Ferreira, T. M. M., B. K. Brisson, A. C. Durham, R. Laufer-Amorim, V. Kristiansen, E. Pure, S. W. Volk, and K. Sorenmo. 2018. "Characteristics of the Epithelial-Mesenchymal Transition in Primary and Paired Metastatic Canine Mammary Carcinomas." *Vet Pathol* 55 (5):622-633. doi: 10.1177/0300985818776054.

[126] Im, K. S., J. H. Kim, N. H. Kim, C. H. Yu, T. Y. Hur, and J. H. Sur. 2012. "Possible role of Snail expression as a prognostic factor in canine mammary neoplasia." *J Comp Pathol* 147 (2-3):121-128. doi: 10.1016/j.jcpa.2011.12.002.

[127] Gamba, C. O., M. A. Rodrigues, D. A. Gomes, A. Estrela-Lima, E. Ferreira, and G. D. Cassali. 2015. "The Relationship Between E-Cadherin and its Transcriptional Repressors in Spontaneously Arising Canine Invasive Micropapillary Mammary Carcinoma." *J Comp Pathol* 153 (4):256-265. doi: 10.1016/j.jcpa.2015.08.006.

[128] Gamba, C. O., K. A. Damasceno, I. C. Ferreira, M. A. Rodrigues, D. A. Gomes, M. R. Alves, R. M. Rocha, A. E. Lima, E. Ferreira, and G. D. Cassali. 2019. "The investigation of transcriptional repression mediated by ZEB2 in canine invasive micropapillary carcinoma in mammary gland." *PLoS One* 14 (1):e0209497. doi: 10.1371/journal.pone.0209497.

[129] Pang, L. Y., A. Cervantes-Arias, R. W. Else, and D. J. Argyle. 2011. "Canine Mammary Cancer Stem Cells are Radio- and Chemo-Resistant and Exhibit an Epithelial-Mesenchymal Transition

Phenotype." *Cancers (Basel)* 3 (2):1744-1762. doi: 10.3390/cancers3021744.

[130] Hollestelle, A., J. K. Peeters, M. Smid, M. Timmermans, L. C. Verhoog, P. J. Westenend, A. A. Heine, A. Chan, A. M. Sieuwerts, E. A. Wiemer, J. G. Klijn, P. J. van der Spek, J. A. Foekens, M. Schutte, M. A. den Bakker, and J. W. Martens. 2013. "Loss of E-cadherin is not a necessity for epithelial to mesenchymal transition in human breast cancer." *Breast Cancer Res Treat* 138 (1):47-57. doi: 10.1007/s10549-013-2415-3.

[131] Timmermans-Sprang, E., R. Collin, A. Henkes, M. Philipsen, and J. A. Mol. 2019. "P-cadherin mutations are associated with high basal Wnt activity and stemness in canine mammary tumor cell lines." *Oncotarget* 10 (31):2930-2946. doi: 10.18632/oncotarget.26873.

[132] Paredes, J., J. Figueiredo, A. Albergaria, P. Oliveira, J. Carvalho, A. S. Ribeiro, J. Caldeira, A. M. Costa, J. Simoes-Correia, M. J. Oliveira, H. Pinheiro, S. S. Pinho, R. Mateus, C. A. Reis, M. Leite, M. S. Fernandes, F. Schmitt, F. Carneiro, C. Figueiredo, C. Oliveira, and R. Seruca. 2012. "Epithelial E- and P-cadherins: role and clinical significance in cancer." *Biochim Biophys Acta* 1826 (2):297-311. doi: 10.1016/j.bbcan.2012.05.002.

[133] Buendia, A. J., C. Penafiel-Verdu, J. A. Navarro, M. Vilafranca, and J. Sanchez. 2014. "N-cadherin expression in feline mammary tumors is associated with a reduced E-cadherin expression and the presence of regional metastasis." *Vet Pathol* 51 (4):755-758. doi: 10.1177/0300985813505115.

[134] Figueira, A. C., C. Gomes, N. Mendes, I. F. Amorim, A. J. F. Matos, P. Dias-Pereira, and F. Gärtner. 2016. "An in vitro and in vivo characterization of the cadherin-catenin adhesion complex in a feline mammary carcinoma cell line." *Clin Diagn Pathol* 1 (1):1-8. doi: 10.15761/NRD.1000103.

INDEX

A

acid, viii, 2, 37, 91
actin cytoskeleton, ix, 7, 20, 37, 67, 68, 105, 108, 110, 112, 113, 114, 130, 131, 133
actinic keratosis, 29
adaptation, 12, 67
adenocarcinoma, 33, 80, 81, 137, 152, 153
adherens junctions, vii, viii, 1, 6, 42, 95, 96, 98, 100, 102, 103, 105, 106, 116, 123, 124, 125, 126, 131, 147, 150
adhesion, vii, viii, ix, 1, 4, 6, 9, 13, 16, 20, 30, 42, 43, 44, 49, 59, 63, 64, 72, 95, 97, 98, 103, 104, 105, 106, 108, 110, 112, 113, 114, 116, 118, 119, 120, 122, 123, 125, 130, 131, 132, 133, 134, 135, 136, 137, 138, 140, 141, 142, 145, 146, 147, 148, 149, 154, 155, 162
adhesive properties, 14, 119
adhesive strength, 105
adiponectin, viii, 2, 7, 12, 27, 44, 45, 46, 58, 61, 62, 63, 64, 65, 71, 72, 86
adipose tissue, viii, 2, 70
aggressiveness, 92, 140, 143
airway hyperresponsiveness, 24
alcohol dependence, 10, 48
alkaline phosphatase, 92
allelic loss, 27
allergen challenge, 25
anatomy, 101, 117
anchorage, 6, 20, 30, 35, 67, 68
anchoring, 98, 103, 108, 111, 124
anchoring junctions, 98, 102, 103, 118
angiogenesis, viii, 2, 13, 14, 16, 17, 21, 34, 39, 44, 45, 64, 65
angioplasty, 18
apoptosis, 13, 22, 23, 30, 35, 63, 65, 135, 141, 143, 149, 150
arrest, 34, 35, 91
arteries, 18, 59, 66
astrocytes, 37, 91
atherosclerosis, viii, 2, 13, 17, 18, 21, 46, 59, 60, 65, 66
avoidance, 106, 121

B

basal cell carcinoma, 2, 29, 73
basal layer, 28, 29, 98
basement membrane, 99, 101, 103

164   *Index*

behaviors, 10, 11, 15, 56, 58
benign, 31, 76, 138, 140
bioavailability, 21, 22
biochemistry, 40, 63, 72, 73, 120
biological psychiatry, 58
bladder cancer, 35, 87
blood, viii, 8, 12, 13, 14, 16, 17, 24, 26, 59, 60, 63, 79, 96, 100, 101, 115, 117, 118, 122, 126
blood flow, 14, 18
blood pressure, 12, 59
blood vessels, 8, 26, 63
blood-testis barrier, viii, 96, 100, 101, 117, 118, 122, 124, 125, 126
breast cancer, 13, 30, 31, 75, 76, 77, 139, 141, 142, 143, 144, 145, 151, 155, 156, 157, 159, 160, 161, 162
breast carcinoma, 76, 136, 143, 154, 157, 160

C

calcium, 44, 69, 104, 119, 131, 133
cancer, viii, x, 2, 13, 27, 30, 31, 32, 33, 34, 35, 36, 37, 39, 43, 76, 77, 78, 80, 83, 84, 86, 88, 130, 132, 134, 135, 136, 137, 139, 141, 142, 144, 145, 146, 149, 152, 155, 161, 162
cancer cells, 31, 34, 77, 88, 141, 142, 144, 146
cancer progression, 27, 30, 31, 34, 35, 39, 78
cancer screening, 36
capillary, 14, 24, 26, 64
carcinogenesis, 72, 85, 86, 139
carcinoma, 2, 3, 4, 27, 29, 31, 34, 36, 37, 64, 73, 74, 76, 80, 81, 82, 83, 85, 86, 87, 88, 89, 90, 130, 136, 137, 143, 144, 151, 152, 154, 155, 157, 160, 161, 162
cardiac muscle, 105
cardiomyopathy, 18

cardiovascular disease, viii, 2, 3, 8, 66, 68
cat, 130, 142, 151, 153
cell biology, 40, 42, 126
cell culture, 150
cell cycle, 13, 19, 34, 35, 84
cell death, 121, 135
cell differentiation, x, 27, 66, 105, 130, 131
cell invasion, 36, 39, 134, 142
cell line, 7, 20, 24, 29, 30, 32, 33, 34, 35, 37, 38, 87, 144, 156, 159, 162
cell membranes, 16, 45
cell metabolism, 135
cell movement, 103, 105, 123, 131, 148
cell surface, viii, ix, 6, 7, 8, 21, 44, 45, 95, 104, 119, 130, 131, 132, 149
cell-cell connections, 96
central nervous system, 121
cerebral cortex, 47
cervical cancer, 36, 89
chemotherapy, 31, 32, 77
chromosome, 5, 27, 39, 76, 93
chronic lymphocytic leukemia, 91
chronic obstructive pulmonary disease, 3, 24, 71
collagen, 7, 19, 90, 148
colorectal cancer, 3, 34, 85, 86
controversial, 100, 109, 140, 145
controversies, 86
coordination, 101
coronary arteries, 18
coronary artery disease, 17
coronary heart disease, 12, 60
correlation, 17, 33, 38, 92, 119, 144, 158
cytokines, 15, 115, 126
cytokinesis, 99
cytomegalovirus, 57
cytoplasm, 6, 104, 136, 140
cytoplasmic tail, ix, 130
cytoskeleton, ix, 6, 7, 16, 20, 37, 42, 67, 68, 98, 103, 105, 107, 110, 112, 113, 114, 117, 125, 130, 131, 133, 143, 146, 149

D

data collection, 140
deficiency, 12, 22, 24, 31, 70
deficit, 2, 10, 52, 53, 58
degradation, 21, 114, 115, 133, 141, 147
desmosome, 103, 104, 107, 108, 109, 110, 124
desmosome-like, 102, 103, 104, 107, 108, 109, 110, 124
detachment, 64, 142
detection, 82, 83, 109
developed countries, 139
developing brain, 10
disease progression, 18, 22, 137
diseases, ix, 12, 40, 63, 66, 130
disorder, 2, 10, 52, 53, 58
diversity, 40, 43, 66, 78, 107, 120, 146
DNA, 16, 20, 32, 36, 75, 77, 79, 80, 81, 82, 83, 89, 92
DNA damage, 83
dogs, vii, x, 130, 136
dopamine, 10
down-regulation, 74, 134, 137, 140, 150

E

E-cadherin, v, ix, 27, 34, 35, 42, 43, 72, 79, 92, 122, 123, 129, 130, 131, 132, 133, 134, 135, 136, 138, 139, 146, 147, 148, 149, 150, 151, 152, 153, 154, 155, 156, 157, 158, 159, 160, 161, 162
ECs, 8, 12, 13, 15, 16, 17, 21, 22, 23, 24, 26, 31, 34, 40, 105, 106
ectoplasmic specialization, 96, 97, 102, 103, 108, 118, 119, 124, 125
emboli, 141, 142
embryogenesis, 8, 39, 134
endometrial carcinoma, 88, 89, 137, 138, 151, 152
endometriosis, 137, 150
endothelial cells, 7, 44, 45, 58, 63, 64, 65, 119, 146
endothelial cells (ECs), 7
endothelial dysfunction, 16
endothelium, 12, 13, 17, 63
environment, viii, 14, 39, 96, 99, 100, 142
environmental factors, 28
epithelial cells, ix, 23, 26, 30, 98, 130, 132, 133, 134, 135, 136, 138, 141, 143, 150
epithelium, viii, 76, 95, 99, 100, 101, 103, 108, 113, 116, 117, 118, 123, 124, 125, 138
esophageal cancer, 33
esophagus, 33, 83
estrogen, 3, 31, 32
evidence, 9, 10, 13, 19, 34, 35, 52, 104, 108, 111, 144
evolution, 5, 40, 86
exocytosis, 26, 132
exposure, 21, 24, 59
external environment, ix, 130
extracellular matrix, 19, 98, 141, 143, 158

F

family members, 5, 6, 105
fertility, iv, vii, ix, 96, 100, 101, 116, 125, 128
fertilization, 118
fibroblast growth factor, 36
formation, ix, 4, 6, 7, 8, 9, 14, 26, 34, 47, 63, 77, 98, 108, 109, 130, 132, 139, 141, 142, 145
free rotation, 105

G

gene expression, 11, 18, 30, 31, 32, 35, 37, 39, 51, 72, 77, 85, 105, 109, 131, 136
gene promoter, 32, 33, 34, 36, 91
gene silencing, 87

gene transfer, 14
genes, 10, 34, 36, 49, 53, 61, 71, 75, 76, 79, 81, 82, 85, 86, 87, 89, 90, 91, 92, 132, 143, 161
genetic alteration, 161
genetic defect, 119
genetic disease, 42
genetics, 48, 49, 52, 53, 60, 62, 79
genome, 3, 10, 48, 49, 52, 60, 62, 71, 78
germ cells, viii, 96, 99, 100, 101, 102, 103, 107, 108, 109, 112, 113, 114, 116, 122, 125
gland, vii, ix, x, 31, 130, 138, 153, 154, 155, 158, 161
glioblastoma, 37, 90
glioma, 37, 91
glucocorticoid, 92
glucose, viii, 2, 22, 25
glucose tolerance, 25
glucose tolerance test, 25
glycoproteins, viii, ix, 95, 104, 130, 131
growth, 3, 4, 6, 9, 13, 15, 29, 30, 31, 32, 33, 35, 38, 40, 47, 65, 69, 70, 74, 75, 90, 91, 97, 105, 106, 107, 131, 142, 151
growth arrest, 38, 40
growth factor, 3, 4, 6, 15, 29, 32, 40, 65, 90, 97

H

hepatocellular carcinoma, 3, 13, 34, 64, 86, 87
hepatocytes, 26, 72
heterogeneity, 28, 138, 140
homeostasis, vii, x, 4, 13, 17, 25, 32, 98, 107, 113, 130
human, x, 3, 8, 12, 13, 18, 26, 30, 35, 38, 40, 43, 45, 46, 49, 51, 53, 57, 59, 60, 61, 62, 63, 64, 69, 73, 75, 76, 77, 79, 80, 83, 84, 85, 86, 87, 89, 90, 91, 92, 106, 119, 123, 130, 136, 137, 138, 139, 141, 143, 144, 146, 150, 160, 162
human brain, 90
human subjects, 12
hyperactivity, 2, 10, 52, 53, 58
hypermethylation, 27, 29, 33, 36, 75, 76, 80, 82, 83, 88, 89, 90, 143
hyperplasia, 24, 36
hypertension, 18, 59
hypertrophy, 14, 23
hypothesis, 9, 108, 136, 138, 145
hypoxia, 14, 135, 141, 159

I

identification, 8, 119, 134, 146, 157
immune response, 16
immune system, 100
immunoglobulin, viii, ix, 95, 130, 131
immunoprecipitation, 8
immunoreactivity, 153
in vitro, 7, 8, 9, 13, 15, 18, 19, 21, 22, 24, 29, 31, 44, 64, 72, 73, 77, 84, 123, 126, 127, 144, 145, 162
in vivo, 7, 8, 12, 13, 14, 15, 18, 21, 22, 29, 30, 31, 64, 73, 77, 83, 123, 126, 127, 145, 148, 162
inflammation, x, 13, 16, 21, 25, 33, 68, 83, 130
inflammatory disease, 70
inflammatory mediators, 90
inflammatory responses, 24
inhibition, 15, 38, 91, 107, 132, 137, 148
injury, iv, 17, 18, 22, 23, 24
insulin, 3, 7, 21, 22, 25, 32, 45, 67, 70
insulin resistance, 21
insulin sensitivity, 21, 25
insulin signaling, 21
integration, 42, 146
integrin, 7, 15, 30, 42, 74, 109, 111, 112, 114, 125, 134

integrins, 107, 112, 150
integrity, viii, 4, 5, 95, 98, 100, 101, 102, 124, 131
intercellular contacts, viii, 2, 20, 74
interface, 6, 103, 108, 115, 132
internalization, 7, 132, 133, 136
intron, 29, 32
invertebrates, 104, 131
ischemia, 14, 23
ischemia-reperfusion injury, 23

K

keratinocyte, 28, 29, 30
keratinocytes, viii, 2, 28, 29, 73
keratosis, 29
kidney, 3, 8, 92

L

laminins, 111
lesions, 18, 25, 29, 31, 34, 38, 76, 89, 137, 145
life sciences, 42
ligand, 7, 20, 45
lipoproteins, 46, 68, 69
liver, 26, 35, 87, 92
liver cirrhosis, 26
localization, 11, 12, 74, 98, 108, 123, 159
loci, 49, 56, 57, 60, 78
locomotor, 11
locus, 32, 33, 59, 61, 76
low-density lipoprotein, 44, 69
lung cancer, 3, 13, 24, 32, 79, 80, 81, 82, 83
lung function, 24, 71
lymph node, 31, 34, 76, 86, 137, 139, 140, 142, 154
lymphangiogenesis, 30
lymphoma, 38, 91

M

majority, 17, 137, 141, 160
malignancy, 137, 145
malignant cells, 142
malignant melanoma, 73
mammary gland, iv, vii, ix, 31, 130, 138, 153, 154, 155, 158, 161
matrix, 19, 30, 67, 74, 108, 116, 141
matrix metalloproteinase, 19
medical, 52, 60, 79, 83, 87
medicine, 53, 60, 84, 88
meta-analysis, 85, 87, 155, 157
metabolic syndrome, 12, 63
metabolism, viii, 2, 26, 69, 88, 159
metabolized, 37
metalloproteinase, 158
metastasis, x, 13, 29, 34, 38, 64, 74, 86, 92, 130, 134, 137, 139, 140, 141, 142, 143, 145, 148, 154, 159, 161, 162
methamphetamine, 10, 49
methylation, 29, 30, 31, 32, 33, 34, 35, 36, 37, 62, 75, 76, 77, 78, 79, 80, 81, 82, 83, 84, 85, 86, 87, 89, 90, 91, 92, 143, 156
mice, 2, 3, 8, 10, 11, 12, 14, 16, 18, 22, 23, 24, 25, 27, 31, 33, 37, 46, 58, 71, 108, 109, 122, 155
microparticles, 3, 14, 16, 65, 66
migration, vii, viii, x, 1, 4, 9, 11, 14, 15, 16, 19, 20, 21, 29, 30, 34, 40, 67, 68, 69, 70, 84, 96, 101, 106, 107, 108, 113, 114, 122, 130, 131, 132, 133, 134, 135, 160
migration routes, 9
mitogen, 3, 4, 69, 97
models, 13, 14, 33, 116, 139, 150
modifications, 113
molecular structure, vii, 2
molecular weight, 7, 65, 98
molecules, vii, viii, 1, 6, 8, 9, 13, 15, 17, 44, 58, 88, 95, 98, 101, 104, 107, 108, 112,

113, 114, 115, 116, 121, 122, 124, 133, 136, 138, 140, 150, 154
monoclonal antibody, 119
monolayer, 14, 19
morphogenesis, vii, ix, 1, 4, 100, 103, 105, 120, 130, 131, 132, 139, 146
morphology, 19, 29, 131, 132
mRNA, 16, 25, 27, 33, 37, 38, 81, 150
mutant, 11, 13
mutant proteins, 11
mutations, 30, 33, 143, 144, 160, 162
myocardial infarction, 23
myocardial ischemia, 23

N

nasopharyngeal carcinoma, 37, 90
N-cadh/catenin complexes, 109
N-cadherin, ix, 9, 27, 96, 104, 105, 107, 109, 111, 114, 115, 123, 149, 157, 162
neovascularization, 14, 23, 141
nephropathy, 61
nervous system, viii, 2, 8, 43, 106
neural connection, 8
neuroblastoma, 37, 90
neurons, 9, 10, 11, 12, 106
neuroprotection, 12
neuroscience, 47, 57
nitric oxide synthase, 3, 127

O

organ, 13, 27, 32, 42, 72, 125
organs, 97, 131, 137, 143
ovarian cancer, 36, 88
oxidative stress, 13, 18, 21, 63, 68
ozone, 24, 71

P

p16INK4A, 80
p53, 36, 70, 155
pancreatic cancer, 35, 87, 147, 148
pathogenesis, viii, 2, 10, 12, 18, 40
pathology, 124, 128
pathophysiological, 8, 10, 16, 34
pathway, 15, 16, 20, 24, 37, 47, 69, 83, 90, 115, 133, 143
pathways, ix, 20, 22, 24, 64, 105, 112, 130, 132, 133, 135
permeability, 24, 115, 126
phenotype, viii, 2, 4, 13, 15, 17, 18, 19, 21, 26, 31, 38, 64, 67, 68, 70, 73, 77, 78, 90, 92, 120, 134, 135, 140, 142, 144
phenotypes, 8, 23, 27, 60, 79, 134
phosphorylation, 16, 21, 22, 29, 34, 36, 112, 113, 114, 115, 126, 133, 147
physiology, x, 24, 66, 80, 116, 130
PI3K, ix, 4, 15, 20, 21, 36, 37, 90, 130
PI3K/AKT, 37, 90
plasma membrane, viii, 2, 5, 6, 16, 22, 103, 136
plasticity, viii, 17, 68, 72, 95, 98, 101, 119, 132, 142, 144, 148, 161
polarity, vii, viii, ix, 1, 4, 9, 32, 78, 95, 98, 105, 106, 107, 109, 121, 122, 126, 130, 131, 132, 134, 136, 148
polarization, 64, 103
polymerase chain reaction, 122
polymorphisms, 12, 34, 59, 61
population, 28, 59, 60, 61, 62, 71, 82, 88, 99, 123
post-transcriptional regulation, 38
pregnancy, x, 130, 136, 150, 151
progenitor cell, 138
progesterone, 38, 136, 150
prognosis, 31, 32, 35, 76, 77, 84, 85, 86, 137, 138, 140, 145, 151, 160

proliferation, viii, x, 2, 4, 13, 15, 16, 18, 19, 20, 28, 29, 32, 35, 37, 39, 46, 65, 67, 69, 74, 84, 90, 100, 105, 130, 131, 133, 135, 140, 152, 154, 158
promoter, 27, 29, 30, 32, 33, 35, 36, 62, 71, 78, 79, 80, 82, 85, 86, 87, 90, 143, 156
prostate cancer, 4, 31, 78, 79
protein kinases, 6, 69, 114, 133, 134
proteins, vii, ix, 2, 6, 12, 16, 17, 19, 30, 39, 44, 45, 46, 58, 98, 102, 103, 104, 105, 106, 107, 108, 110, 111, 112, 114, 115, 119, 130, 131, 134
proteolysis, 104, 113
proto-oncogene, 20, 97
prototype, 27, 106
psoriasis, 30, 75
psychiatry, 48, 49, 56, 57
psychopharmacology, 58

R

receptor, 3, 7, 15, 20, 21, 23, 24, 29, 31, 32, 45, 63, 64, 66, 70, 79, 97, 112, 114, 115, 126
reciprocal relationships, 158
recognition, 4, 8, 104, 106, 107, 131
recurrence, 79, 81, 87, 137
recycling, 113, 114, 115, 132, 134, 147
regeneration, 21, 27, 28, 46, 72
relevance, viii, 2, 8, 20, 24, 44, 63, 139, 145
remodelling, 101, 113, 114, 115, 133, 147
repression, 36, 37, 88, 143, 161
requirement, 15, 26, 64
residues, 104, 114
resistance, 21, 68, 106, 135, 141, 143
response, 10, 14, 15, 16, 17, 19, 23, 24, 25, 31, 32, 37, 44, 50, 60, 65, 67, 71, 77, 83, 159
responsiveness, 21, 24, 32
restenosis, 13, 18, 21, 63
reticulum, 4, 12, 13, 15, 65, 103

risk, 10, 11, 20, 29, 34, 35, 53, 56, 58, 59, 60, 70, 76, 79, 82, 83, 86, 88, 156, 159
risk assessment, 82
risk factors, 83, 156
RNA, 4, 15, 37, 90

S

schizophrenia, 11, 57
science, 77, 82, 83, 87, 116
secretion, 25, 26, 70
seminiferous tubules, iv, v, vii, viii, 95, 96, 98, 99, 100, 101, 102, 103, 104, 106, 107, 108, 109, 111, 113, 114, 115
sensitivity, 22, 35, 36, 62, 149
sertoli cells, viii, 96, 99, 100, 101, 102, 103, 110, 111, 113, 114, 117, 118, 124
Sertoli cells, viii, 96, 99, 100, 101, 102, 103, 110, 111, 113, 114, 117, 118, 124
serum, 11, 19, 23, 25, 32, 35, 36, 58, 79, 80, 89
signal transduction, ix, 6, 15, 20, 40, 105, 107, 130
signaling pathway, vii, 1, 30, 34, 36, 37, 70, 84, 117, 127, 135
signalling, ix, 42, 45, 69, 90, 105, 106, 107, 115, 130, 131, 135, 146
skin, 28, 29, 30, 73
skin cancer, 29
smooth muscle, viii, 2, 4, 8, 17, 18, 44, 45, 66, 67, 68, 69, 70
smooth muscle cells, viii, 2, 8, 44, 45, 67, 68, 69, 70
somatic cell, viii, 96, 99, 101
somatic mutations, 39
specialization, 96, 97, 103, 118, 119, 125
species, viii, 5, 20, 95, 99, 100, 104, 107, 109, 111, 139
spermatid, ix, 96, 103, 109, 124

spermatogenesis, 96, 98, 99, 100, 101, 103, 107, 112, 115, 116, 117, 118, 121, 123, 125, 126, 127, 148
spermatogonial stem cells, 108, 123
spermatogonium, 97, 99
squamous cell, 4, 29, 73, 74, 83, 90
squamous cell carcinoma, 4, 29, 73, 74, 83, 90
stability, 105, 131, 132
stabilization, 36, 70, 108, 142
stress, 7, 12, 13, 15, 17, 18, 19, 21, 23, 58, 65, 108
stress response, 12
structural protein, 114, 115, 133
structure, vii, viii, 2, 6, 43, 96, 98, 103, 105, 106, 113, 116, 131
survival, vii, x, 1, 4, 15, 16, 19, 22, 31, 33, 34, 36, 37, 44, 58, 67, 82, 84, 90, 100, 108, 130, 137, 140, 141, 142, 149, 154, 155, 156, 157, 158
synthesis, 12, 20, 26, 113, 132

T

target, 3, 8, 17, 22, 37, 39, 82, 125
testis, viii, 95, 96, 98, 99, 100, 101, 102, 103, 107, 108, 109, 110, 111, 115, 117, 118, 119, 122, 123, 124, 125, 126, 127, 148
testosterone, 115
tissue, vii, viii, ix, 1, 4, 5, 8, 9, 12, 14, 18, 20, 22, 23, 27, 28, 30, 31, 32, 35, 36, 46, 95, 97, 99, 100, 103, 105, 106, 107, 120, 121, 130, 131, 132, 141, 142, 145, 146, 147, 159, 160
tissue homeostasis, 131
trafficking, 26, 74, 113, 114, 134
transcription, 10, 13, 19, 30, 36, 67, 87, 132, 135, 141, 143, 144
transcription factors, 19, 67, 135, 143, 144
transformation, 27, 29, 30, 34, 37, 73, 141

transforming growth factor, 19, 97
transitional cell carcinoma, 88
translocation, ix, 20, 90, 96, 100, 137
tubulobulbar complexes, 97, 103
tumor, 3, 13, 14, 27, 28, 29, 30, 31, 32, 33, 34, 36, 37, 38, 40, 44, 63, 64, 77, 80, 81, 84, 87, 88, 90, 91, 149, 151, 152, 156, 159, 162
tumor cells, 64, 87
tumor development, 14
tumor growth, 29, 30, 31, 37
tumor metastasis, 149
tumor progression, 32, 36, 64, 80, 149
tumorigenesis, 78
tumour growth, 141, 144
turnover, 20, 28, 112, 132, 147
type 1 diabetes, 61
type 2 diabetes, 61
tyrosine, 69, 97, 112, 114, 126

U

ulcerative colitis, 85
uterus, iv, vii, ix, x, 130, 138, 150, 151, 153

V

vascular diseases, 68
vascular endothelial growth factor, 4, 84, 155
vascularization, 14, 31
vasculature, 8, 25, 44, 59
vertebrates, 5, 98, 104, 106, 131
violent behavior, 11, 56
vulnerability, 49, 70

W

Wiskott-Aldrich syndrome, 96, 97, 111, 125

Wnt signaling, 120
working memory, 53